Roadside Geology
of NEBRASKA

Harmon D. Maher Jr.,
George F. Engelmann,
and Robert D. Shuster

2003
MOUNTAIN PRESS PUBLISHING COMPANY
Missoula, Montana

Cover image from
U.S. Geological Survey EROS Data Center
See explanation of landscape features on pages 2 and 3.

Roadside Geology is a registered trademark
of Mountain Press Publishing Company.

Library of Congress Cataloging-in-Publication data
Maher, Harmon D., 1955-
Roadside geology of Nebraska / Harmon D. Maher, Jr., George
F. Engelmann, and Robert D. Shuster.
      p. cm.
Includes bibliographical references and index.
   ISBN 0-87842-457-1 (pbk. : alk. paper)
1. Geology—Nebraska—Guidebooks. 2. Nebraska—Guide-
books. I. Engelmann, George Felix, 1950- II. Shuster, Robert
Duncan. III. Title.
   QE135 .M34 2003
   557.82—dc21

                                                   2002010930

Printed in the United States of America

Mountain Press Publishing Company
P.O. Box 2399 • Missoula, MT 59806
(406) 728-1900

Crossing the craton—the Stable Interior Craton, core of the continent: Illinois, Iowa, Nebraska, and friends—you don't see a lot of rock. . . . seldom does it outcrop, and, where it does, it is such an event that it is likely to have been named a state park.

—John McPhee
From essay "Crossing the Craton" in
Annals of the Former World

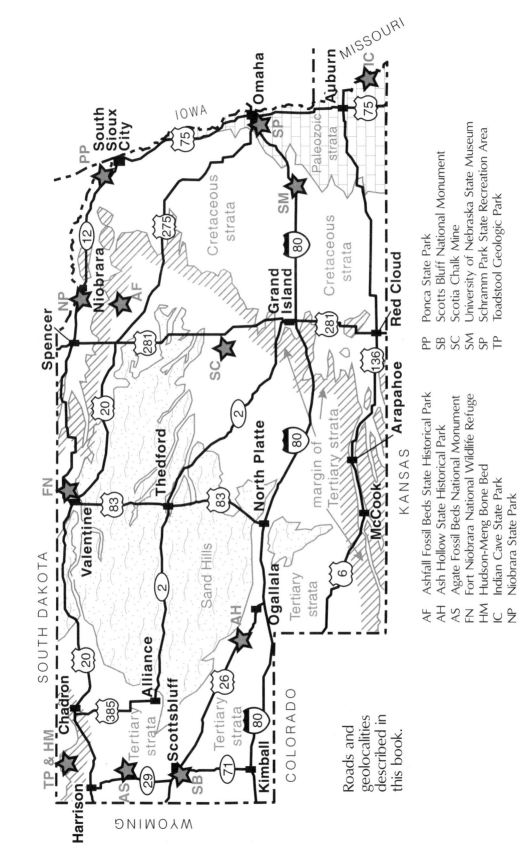

Roads and geolocalities described in this book.

AF   Ashfall Fossil Beds State Historical Park
AH   Ash Hollow State Historical Park
AS   Agate Fossil Beds National Monument
FN   Fort Niobrara National Wildlife Refuge
HM   Hudson-Meng Bone Bed
IC   Indian Cave State Park
NP   Niobrara State Park

PP   Ponca State Park
SB   Scotts Bluff National Monument
SC   Scotia Chalk Mine
SM   University of Nebraska State Museum
SP   Schramm Park State Recreation Area
TP   Toadstool Geologic Park

CONTENTS

Road Guides

ACKNOWLEDGMENTS

In writing this book we must acknowledge our debt to the many geologists and paleontologists who have researched and published on the geology of Nebraska. Our work consists largely of a distillation and reorganization of that large body of work, written in terms that we hope the layperson finds understandable and interesting. We have relied particularly heavily on the work of geologists at the University of Nebraska Department of Geology, State Museum, and Nebraska Conservation and Survey Division in Lincoln, and the American Museum of Natural History in New York. We have particularly benefited from the depth and breadth of work on the geology of Nebraska by Marv Carlson, Bob Diffendal, Ted Galusha, Bob Hunt, Roger Pabian, Jim Swinehart, and Mike Voorhies. A special acknowledgment is due to the lifetime of work by Morris Skinner whose observations and collections supported not only his own research on the geology and paleontology of Nebraska, but that of many others who have followed or who are yet to come. New geoscientists continue to become involved in this grand puzzle and the research continues; we and this book have benefited from their efforts to reveal Nebraska's fascinating geology. Hannon LaGarry, Roger Pabian, and Jim Swinehart read through an early draft of the book and we greatly appreciate their comments and insights. We also thank Rolfe Mandel for providing feedback on part of the book, and for his enthusiastic willingness to answer questions and share his expertise on things Quaternary. Naturally, any mistakes herein are the authors' alone. We would also like to thank Jennifer Carey for her careful and skillful editing.

This book was, in part, born out of field trips with students, so we also thank those students who wanted to learn about Nebraska's geology, and who were willing to put up with long drives, wind and rain storms, and sleeping in tents in order to do so. We thank these students for their enthusiasm, their questions, and their good humor. Out on the road and in the field are the best places to experience geology, but it does demand some additional effort and sacrifices.

Finally, the authors would also like to thank their families for their willingness to put up with our absence at times. Harmon would also like to thank his family for their willingness to travel the roads of Nebraska with him at other times. We greatly appreciate our families' significant support as we worked on this book.

PREFACE

In the 1800s, settlers and scientists alike rushed across the plains to the mountains to make their fortune and fame in the West, leaving mostly ruts on the Nebraska landscape. Nowadays, many people travel I-80, following the Platte River valley across Nebraska on their way to or from the Rockies, where the geology and landscape are readily apparent, in places extravagantly so. Such travelers might expect that a book on the geology of Nebraska would be quite short—simply describing sandy and muddy rivers crossing flat plains. But Nebraska has many remarkable geologic features: the largest dune field in the Western Hemisphere, an unsurpassed fossil record of the evolution of the diverse mammals that roamed these plains throughout the later part of the Age of Mammals, sedimentary rocks that record of the rising of the Rocky Mountains to the west, and a landscape of canyons, bluffs, and badlands. To see Nebraska's unique geologic features, you will need to get off the thruway and explore the countryside. Take *Roadside Geology of Nebraska* with you as a guide to the landscape, the rocks, and the geologic history.

Nebraska's geology can be particularly subtle, not apparent to the untrained eye. The significance of this or that deposit is not immediately obvious. A simple gravel deposit is much more interesting when you discover that unusual rocks in it point to a far away source, and when you find out why it is on the top of a ridge and not down in a valley. A few bones poking from a streambank may be part of a bone bed with hundreds of individuals, and you may wonder how they came to die together. While subtle in its manifestation, Nebraska's geology reveals a dramatic history and those who venture to unlock that history come to appreciate it.

This guide begins with an introduction to geologic time and sedimentary rocks for the nongeologist. We then provide a geologic framework for Nebraska with general discussions about what happened to this region through geologic time. Following the overview, road guides describe the geology along a suite of highways, many of which follow rivers. We chose to follow these roads because they are common directions of travel, and rivers provide good places to see the geology.

Scattered throughout the road guides are geologic points of interest, which we call geolocalities. Many lie off the beaten path and are side trips from the main highway. We include geolocalities for those who have some extra time and want to get out of the car and see geology up close. Nebraska is rich in such sites, and we strongly encourage the traveler to take the time to see them. These sites may be particularly useful to educators, especially those planning field trips. Stars within the text and on the maps identify the geolocalities.

Reading Geologic Maps

Throughout this book are a series of simplified geologic maps. A map that shows exactly what you can see at the surface would not be useful for depicting geology. It would consist of a whole bunch of tiny dots or strips that represent outcrops, with large areas of vegetation or other cover in between. A geologic map is usually an interpretation of what the surface would look like if you were to strip away the cover.

Geologic maps come in a variety of types. Some maps have very specific purposes, showing, for example, the paths that ancient rivers followed, the distribution of structures in the earth's crust, or the position of distinctive landforms. Soil maps depict the types of soil present at the surface. Maps of recent or surficial geologic deposits, such as unconsolidated alluvium, loess, or till of Quaternary age, show what you would see if you removed the vegetation and soil. Bedrock maps attempt to show what hard rocks would be exposed if you stripped away vegetation, soil, and unconsolidated material. If a certain rock depicted on the map is not apparent at some specific locality, you may just have to dig to get to it. With Nebraska's wealth of Quaternary deposits you may have to dig down hundreds of feet before you get to the bedrock.

The maps in this book are a peculiar hybrid of bedrock, surficial cover, and landform maps. They highlight the more obvious, interesting, and sometimes diverse geologic features that you can see along the various roads. Surficial deposits strongly dominate the landscape along some routes and so we emphasized them. Along other routes, you can see more of the underlying bedrock.

Sometimes smaller exposures of a geologic unit, perhaps those capping some solitary buttes, may be too small to include on a map where the scale is generally an inch to miles. Relatively thin but important geologic units are also difficult to show on a map. Even a thin map line may be many times wider than an accurately scaled representation. It may be helpful to think of these maps as somewhat impressionistic renditions of the actual geology, correct in broad stroke but by their nature unable to capture some level of detail. The U.S. Geological Survey and the Conservation and Survey Division of the University of Nebraska publish a variety of more detailed geologic maps, with a wealth of information on them for those who wish to pursue these subjects at the next level of detail.

Patterns Used on Maps for Major Rock Units in Nebraska

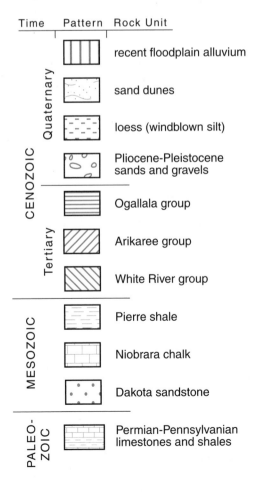

Time	Pattern	Rock Unit
CENOZOIC — Quaternary		recent floodplain alluvium
		sand dunes
		loess (windblown silt)
		Pliocene-Pleistocene sands and gravels
CENOZOIC — Tertiary		Ogallala group
		Arikaree group
		White River group
MESOZOIC		Pierre shale
		Niobrara chalk
		Dakota sandstone
PALEOZOIC		Permian-Pennsylvanian limestones and shales

Shaded-Relief Maps: We've included a handful of shaded-relief maps constructed using digital elevation models, known as DEMs, to show subtle landscape features. DEMs are gridded descriptions of the land's elevation. A known elevation value is used for each grid intersection. A 30-meter DEM means that the grid has a 30-meter spacing. From this array of elevations, scientists use computers to construct shaded-relief maps. In these maps darker areas are lower in elevation and lighter areas are higher. For instance, a river valley will be darker than the surrounding uplands. DEMs look a little bit like air photos but simpler. If the soil, rock, and vegetation cover of the land surface was totally uniform and you photographed it under certain lighting, then the photo would look like a shaded-relief map. Most human developments such as roads and buildings are not on DEMs, though larger features, such as an airport runway or a railroad grade can be. Such simplified images can reveal very detailed and subtle changes in the landscape.

We provide some references for further information, and the Conservation and Survey Division of the University of Nebraska offers a number of very helpful publications for those who want to learn more. The field of geology has broadened in the last few decades, with new interests and responsibilities, and in addition to more traditional descriptions of the geology, we have mentioned geologic-related environmental concerns in Nebraska. We have also included air photos to help you see the geology from a different perspective, one particularly useful for Nebraska. These air photos are from a large U.S. Geological Survey repository available through the TerraServer website.

Enjoy your travels and we hope this book will help enhance your appreciation of Nebraska's geology.

GEOLOGIC TIME SCALE

Era	Period	Epoch	million years ago
CENOZOIC	Quaternary	Holocene	.01
		Pleistocene	1.6
	Tertiary	Pliocene	5
		Miocene	24
		Oligocene	36
		Eocene	58
		Paleocene	65
MESOZOIC	Cretaceous		145
	Jurassic		208
	Triassic		248
PALEOZOIC	Permian		286
	Pennsylvanian		320
	Mississippian		360
	Devonian		417
	Silurian		443
	Ordovician		495
	Cambrian		542
PRECAMBRIAN	Proterozoic Eon		2,500
	Archean Eon		

Geologic Events in Nebraska

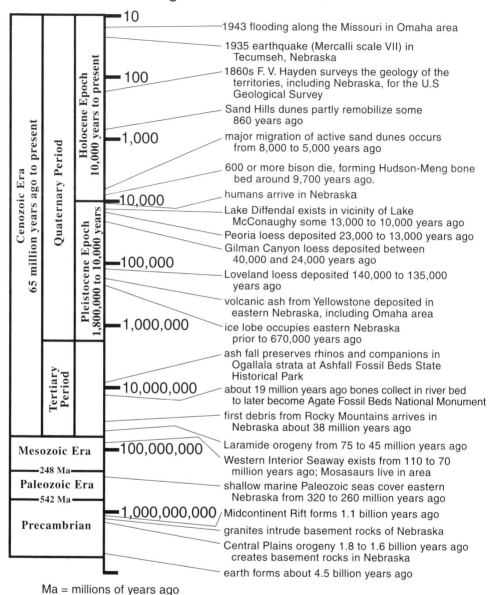

Ma = millions of years ago

Events in the geologic history of Nebraska with years before present shown on a logarithmic scale. The farther back in time we look, the less clearly we see, so geologists know more about recent history than older history. The recent end of a typical time line of known geologic events gets impossibly crowded. A logarithmic scale provides the appropriate room to list recent geologic events. It also reminds us of our time-related myopia.

Geologic Summary
of NEBRASKA

Mountains are not evident in the plains of Nebraska now, but buried beneath a thick blanket of sedimentary rocks are the eroded roots of mountains that grew upward more than one and a half billion years ago in the Precambrian era. The roots consist of metamorphic gneisses and igneous rocks that were injected, molded, and heated deep in the interior of the mountains. Some 1.4 billion years ago unusual granites intruded these rocks, and 1.1 billion years ago basalts were extruded on the surface in southeast Nebraska as the North American continent almost rifted apart. These rocks serve as a foundation of sorts, so they are known as basement rocks.

The sedimentary rocks exposed at the surface in Nebraska tell us stories of the rise and fall of distant mountains, the invasion and retreat of seas and continental glaciers, and of extensive storms of dust, sand, and sometimes volcanic ash. Many times during Paleozoic time, shallow seas submerged the continental interior, leaving behind limestones, sandstones, and shales, with a record of ancient, mainly marine life entombed within. Rivers draining the newly formed Appalachian Mountains to the east transported sediment to the sea's eastern shoreline, which at times was in eastern Nebraska. Another sea, about 110 to 70 million years ago in Cretaceous time, flooded the continental interior and much and sometimes all of Nebraska. It left behind extensive deposits of sand, shale, and chalk with fossils of more advanced life forms such as large sea lizards.

The Rocky Mountains began to rise to the west as the Cretaceous sea receded, and by 38 million years ago sediment shed from the Rockies had spread into Nebraska. Sediment continued to build up until some 2 million years ago, reaching three-quarters of the

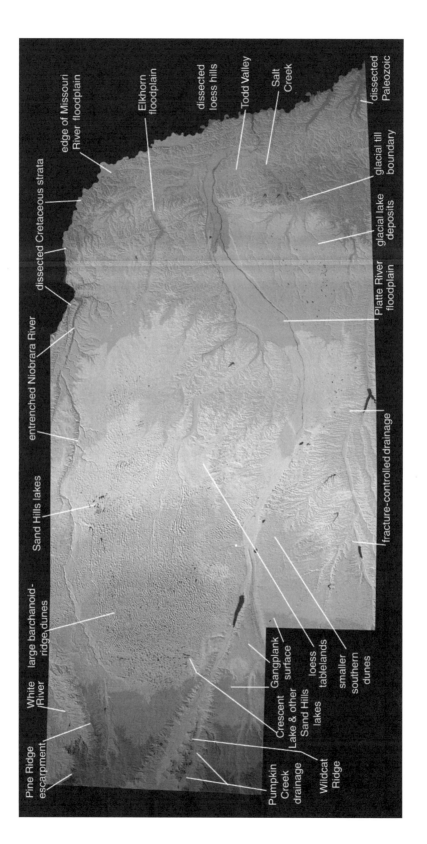

Pine Ridge escarpment

White River

large barchanoid-ridge dunes

Sand Hills lakes

dissected Cretaceous strata

edge of Missouri River floodplain

Elkhorn floodplain

dissected loess hills

Todd Valley

Salt Creek

dissected Paleozoic

glacial till boundary

glacial lake deposits

entrenched Niobrara River

Platte River floodplain

fracture-controlled drainage

Pumpkin Creek drainage

Wildcat Ridge

Crescent Lake & other Sand Hills lakes

Gangplank surface

loess tablelands

smaller southern dunes

← **OPPOSITE:** Several major geologic features are evident on this shaded-relief map made from a digital elevation model (DEM) of Nebraska. The boundaries between different shades of gray follow elevation contour lines, with a 500-foot interval. In eastern Nebraska, the elevation is about 1,000 feet above sea level, and it climbs steadily to nearly 5,000 feet in the west. This gently sloping surface across the western three-quarters of Nebraska is due primarily to the wedge of Tertiary sediment that built out from the Rocky Mountains. Erosion and deposition have significantly modified this surface.

The Sand Hills are most conspicuous where the barchanoid-ridge dunes are well formed. Eolian sands actually cover a larger area than evident here, especially to the east, where the dunes are smaller and more variable in form and so don't show up as well on the DEM. Note the clusters of lakes (dark points) that occur in some parts of the Sand Hills.

The valleys of the Platte, Republican, Loup, and Elkhorn Rivers show well-developed flat floodplains and extensive drainage development. In contrast, the Niobrara River is entrenched, with a narrower and deeper valley.

You can see the western margin of glacial moraines, especially south of the Platte River—a much more intricate drainage pattern developed on the tills to the east. The relative impermeability of the tills may have led to greater runoff and thus dissection. In contrast, in the more porous sediments to the west, more of the rainfall soaks into the ground. In the west, the dominance of erosion is reflected in the various escarpments and ridges associated with the White, Niobrara, and North Platte Rivers and their tributaries. The Gangplank surface they have cut into is evident as the relatively smooth surface that slopes up to the west.

way across Nebraska. During much of this time Nebraska resembled a gently east-sloping African savanna, but with a changing and exotic array of saber-toothed cats, three-toed horses, giraffe-necked camels, small rhinos, and other beasts.

Then, during the Pleistocene ice ages, starting a bit less than 2 million years ago, glaciers rearranged the landscape, diverting rivers to the south, forming lakes, and depositing debris in eastern Nebraska. The Missouri River developed in the path of a retreating glacial lobe. At times when the glaciers were farther north, winds blew dust across the plains. Grasslands trapped the dust, and built upward, accumulating locally thick loess deposits that mantled the older landscape.

Then, around 10,000 years ago and later, with the disappearance of the North American ice cap, much of central Nebraska became a stark sea of blowing and drifting sand—part of the largest dune field in the Western Hemisphere. Humans formed camps on edges of lakes associated with the dunes some 10,000 years ago and possibly were in the area even earlier. As the climate changed and rainfall increased, grass covered and stabilized the dunes. The dunes marched again during intervals of decreased rainfall, most recently 860 years ago.

We'll introduce Nebraska's geologic history in more depth, but first we want to discuss the nature of geologic time, fossils, and sedimentary rocks—important background information for understanding Nebraska's geology.

Geologic Time

An appreciation of the enormity of geologic time is critical to understanding geology and the evolution of life. Charles Darwin recognized this in 1859 in *On the Origin of Species* when he wrote, "He who . . . does not admit how incomprehensibly vast have been the past periods of time, may at once close this volume." Even the relatively limited geology in Nebraska records such a long series of disparate events in an incomplete and sporadic geologic record that one must view it within the framework of geologic time.

The oldest rocks exposed in Nebraska, in the eastern part of the state, are 300- to 250-million-year-old sedimentary strata. Much older Precambrian rocks, the basement rocks of Nebraska, are deeply buried beneath the cover of sedimentary rocks thousands of feet thick. Geologists have sampled the basement rocks by drilling; among these rocks are distinctive granites dated at 1.4 billion years old. Still, even these granites are only one-third the age of the earth, which is about 4.6 billion years old. The immensity of geologic time, so overwhelming in comparison to the ephemeral human time frame and human history, has sometimes been referred to poetically as "deep time," and it challenges the human psyche. The last 300 or so million years represented in Nebraska's surface geology is only some 8 percent of total earth history, but it is long enough nonetheless.

Since the beginnings of scientific geology about 200 years ago, geologists have been constructing and refining a geologic time scale. The scale divides history into intervals, each of which has a name. This approach is somewhat familiar to people who study human history or literature. Think of the Dark Ages or the Elizabethan period. The largest geologic units are eras, which are broken up into periods, which in turn are subdivided into epochs.

This time scale is based primarily on a unique sequence of fossils in the sedimentary rock record that was dictated by evolution and extinction. Sediments entombed the animals or plants, creating a record of what lived at that time, and certain life forms lived in certain geologic periods. If fossil form A is always found exclusively in strata underlying, and thus older than, sediment containing fossil form B, and the two are not found together, one can deduce that A lived and went extinct before B lived. A unit of time within the geologic time scale is based on the temporal information from a large host of fossils. The time scale is a common reference framework for geologists who reconstruct events in earth history and place new findings into the framework on the basis of their fossil content, stratigraphic position, age as determined by radioactive dating, or by other unique characteristics.

The boundaries of the four eras—Precambrian, Paleozoic, Mesozoic, and Cenozoic—were determined by major developments or changes in life forms. The beginning of Paleozoic time marks the first widespread presence of marine organisms with skeletons. The Paleozoic-Mesozoic boundary and Mesozoic-Cenozoic boundary mark major extinction events—for example, we find non-avian dinosaur fossils only in rocks of Mesozoic age, not in rocks of Cenozoic age.

Geologists also use radioactive material to date rocks. Some minerals bear radioactive isotopes that decay at a constant rate, which permits geologists to assign an age in years to geologic events in the time scale—a science known as geochronology. The minerals in igneous rocks are especially useful for dating purposes. Volcanic ashes are common in certain parts of Nebraska, and these can sometimes be dated on the basis of the radioactive material in minerals they contain.

Understanding Sedimentary Rocks

All of the exposed rock units in Nebraska are sedimentary—deposited by wind, water, glaciers, and other transport agents on the earth's surface. The only possible exception might be various volcanic ash layers, which are igneous in composition but accumulate like sedimentary rocks. Volcanic ash straddles the division between sedimentary and igneous rocks. Though originally formed from molten material, volcanic ash is carried by the wind and often

grain size diameter in millimeters (mm)	sediment type	sedimentary rock
larger than 2 mm	gravel	conglomerate
1 to 2 mm	very coarse sand	very coarse sandstone
$1/2$ to 1 mm	coarse sand	coarse sandstone
$1/4$ to $1/2$ mm	medium sand	medium sandstone
$1/8$ to $1/4$ mm	fine sand	fine sandstone
$1/16$ to $1/8$ mm	very fine sand	fine sandstone
$1/256$ to $1/16$ mm	silt	siltstone
smaller than $1/256$ mm	clay	mudstone, claystone, shale

Sediment and sedimentary rock classification based on grain size.

modified by subsequent surface processes, such as burrowing by organisms. Thus, ash layers have distinct sedimentary features and textures.

Sedimentary rocks are classified on the basis of particle size, composition, fossil content, internal structures and textures, and in some cases by how they were transported to their present position. Basic terms that describe grain size—such as sand, silt, and clay—are often modified with adjectives that further describe coloration or composition. For example, a sedimentary rock could be described as a brown, fossiliferous, fine-grained sandstone.

Grain size, texture, composition, and a variety of structures and fossils in sediments help geologists deduce how sedimentary rocks developed. The grain size of a sediment often gives information as to how the sediment was transported and deposited. For example, normal terrestrial winds cannot move gravel but fairly strong water currents can. Glaciers carry all sizes of debris and often dump them together in a deposit containing everything from boulders to mud.

The composition of a sediment tells us something about the source terrane, the area where the sediment originated. From Cenozoic time through present, the Black Hills, Laramie Mountains, and Colorado Front Range have shed cobbles of metamorphic and igneous rocks from the interior of these uplifts into western Nebraska. Some cobbles in Nebraska gravels are so distinctive that geologists have traced them back to the specific rock body in the mountains they eroded from.

Sediments are modified after deposition by burrowing organisms, soil development, and the precipitation of minerals from the groundwater, sometimes forming whimsical or even humorously shaped concretions—for example, the pipelike concretions at Scotts Bluff. Over time, cementation and compaction turn the sediment into sedimentary rock.

When geologists describe the stratigraphic history of a region, they often attempt to reconstruct the depositional environment— the landscape and climate of the time. A guiding principle that helps in this endeavor is uniformitarianism—the notion that the processes in operation on earth today are the same ones that operated in the past. Geologists have debated this concept a great deal since it was initially developed in the early 1800s. To understand the concept, imagine a geologist who thinks she recognizes evidence of past sand dunes in sedimentary rock. To corroborate her findings, she observes the formation and movement of active sand dunes. When comparing sand dunes to other modern-day depositional environments, she finds that some traits are unique to sand dunes. For example, vertebrate or insect tracks on large sloping sand surfaces where the sand is fine only occurs in sand dunes. When she finds the same unique traits in a crossbedded sandstone, then she can infer that sand dunes existed there in the geologic past. In another example, the study of ash layers produced by modern volcanoes enabled geologists to recognize older, chalky white layers as volcanic ash in sedimentary rocks.

As the saying goes, "the present is the key to the past." Using this basic approach to help reconstruct Nebraska's past, geologists have discovered that Nebraska has hosted warm, shallow marine seas, grassy plains inhabited by large and exotic animals, glaciers, and sand dunes.

Defining Rock Units

In order to communicate with each other, geologists divide the sequence of sedimentary rocks into stratigraphic units. A systematic set of rules, known as the stratigraphic code, governs the naming and description of a rock unit. The location where a distinctive unit is first described is the *type locality* and usually gives the unit its name: the Valentine formation was first defined on the basis of strata near Valentine, Nebraska. Type localities of various formations dot the Nebraska landscape. There is a hierarchy of stratigraphic units from group to formation to member to bed, largest to smallest. For example, the Orella member is within the Brule formation, which is within the White River group. These stratigraphic names are like place names in geography, markers in the stratigraphic world.

Each unit includes strata distinguished by some suite of similarities in the rock; its boundaries are where some change caused deposition of a different type of sediment. Color, texture, composition, and fossil content are all used to distinguish stratigraphic units.

The modern landscape often indicates the presence or absence of a particular rock type. Some stratigraphic units form cliffs while others form slopes because some units are more resistant to erosion than others, a phenomenon known as differential erosion. A good example in western Nebraska is the Arikaree group strata, which tend to form cliffs, bluffs, and ridges. At the foot of some bluffs are badlands that tend to form in White River group strata.

Gaps in the Record

Early in the study of geology, scientists recognized the incomplete character of the rock record. *Unconformities* are depositional contacts between rock units where there is a large gap in the geologic record. Different types of unconformities exist. Where the underlying and older rock is granite and metamorphic rocks it is a *nonconformity*. The contact in the subsurface between Nebraska's basement rocks and the overlying Paleozoic sediments is a nonconformity. Where the underlying rock is tilted, eroded strata at an angle to the overlying strata, then it is an *angular unconformity.* A more subtle type of unconformity, a *disconformity,* is the most common in Nebraska and occurs where there is a gap in the record, but the strata above and below appear to be parallel to each other. Unconformities are caused by periods of uplift and erosion or nondeposition, which in turn can be caused by episodes of crustal deformation or changes in sea level. An unconformity between

Cretaceous and Tertiary sediments in Nebraska is associated with the Laramide orogeny, an important mountain-building episode some 75 to 45 million years ago that formed not only the Laramie Mountains in eastern Wyoming, but many other major mountain blocks in Wyoming and Colorado. These uplifts shed sedimentary aprons to the east that eventually reached into Nebraska.

Periods of long landscape stability, when nothing is deposited and erosion is inappreciable, can also be associated with unconformities. Soil usually forms during these periods of stability and if deposition resumes at some later date, then the soil may be buried and preserved. Ancient soils, called paleosols, are common in some parts of Nebraska's rock record and suggest periods of relative stability.

During the period of surface stability, chemical weathering and biologic processes act for a prolonged time on the material near the surface to form soils. Plants add organic matter to the upper portion. Water that seeps through the soil material can alter minerals, for example feldspar to clays. Soluble material, such as calcium carbonate, can be dissolved and sometimes reprecipitated in a middle zone. Iron and other oxides can form in the soil. The results of this host of processes are visibly different zones of altered material that extend from the surface down into the unmodified rock or sediment below. Soil profiles are typically a few feet thick but can extend hundreds of feet in cases of intense weathering and prolonged exposure. Soil characteristic of arid climates, in which calcium carbonate collects in the middle soil horizon, is commonly preserved in the geologic record in Nebraska. Buried soils are important for recognizing different loess units of Quaternary age in Nebraska.

While ancient, buried soils are sometimes found in the material below an unconformity, they are also found within stratigraphic units. Here they represent smaller gaps in the depositional record, perhaps thousands to tens of thousands of years, and not major unconformities. Soils *within* units reminds us of the very episodic nature of deposition.

Longer gaps can be used as convenient boundaries between stratigraphic units. Nebraska has three major gaps in its geologic record, the longest of which is perhaps 1 billion years, separating rocks of Precambrian time from rocks of Paleozoic time.

Does the nature and imperfection of the geologic record preserve some events more than others? Do the small-magnitude, very frequent, day-to-day events such as shifting sandbars or the large-magnitude, infrequent, catastrophic events such as volcanic eruptions shape and bias the record more? Nebraska's sediments indicate that the answer is a mix. However, in the past many geologists have underestimated the effect of the large-magnitude events.

Fossil Formation

Of all the objects of geological study, fossils probably have the greatest power to stir the imagination. Here in Nebraska, one of the most important aspects of the geology is its remarkable fossil record. Fossils can be simply defined as evidence of prehistoric life—the direct record of an organism in the environment. Fossils can be tracks or other evidence of an organism's passage, or the remains of the organism itself. Stone tools and other objects manufactured by humans, even in prehistoric times, are not called fossils, but artifacts.

Preservation creates a fossil from some relic of life and can be accomplished in many ways. Trace fossils, like tracks, are preserved by the same processes that turn sediments into rock. It is only necessary that these ephemeral features be buried rapidly before surface processes obliterate them. Rapid burial is usually the key to preserving the remains of the organisms as well. As long as the body of an animal is exposed on the surface, whether on land or on the seafloor, scavengers will eat what they can and scatter the rest. Flooding rivers may sweep away the remains and tumble them along the streambed before burial. This is why isolated pieces of an animal, like teeth and bones, are much more common than complete skeletons, which only survive if the animal was buried soon after death. Even more rare are those places where many animals were buried virtually undisturbed, just as they were when they died, usually the result of some catastrophic event. The Ash Fall fossil beds and the Hudson-Meng bone bed are two such exceptional localities in Nebraska. Although the remains of plants respond to the environment in different ways from those of animals, the same kinds of processes affect them.

Buried or not, organic material tends to decompose and return to the environment. Fortunately for paleontologists, many creatures construct a skeleton or other hard structures that incorporate minerals, making them more durable. The bones of the fossil mammals from Cenozoic time, or the shells of mollusks from the Pennsylvanian and Cretaceous seas are examples of such structures.

There is a popular misconception that fossils have been "turned to stone," but in most cases this is not so. The organic material that was once part of the structure usually decomposes quickly, making the bone or shell much less resilient, and the space that was once occupied by the organic matter may become filled to a greater or lesser extent by minerals deposited from the groundwater, making it heavier. But, for the most part, the bones and shells still contain the same mineral matter deposited by the animal as it grew. In some cases, the groundwater does dissolve the original minerals as it replaces them with others so that only the form of the shell is preserved. Plants can be preserved in this way also, as in some ancient soil horizons in Nebraska that contain roots and seeds replaced by calcium carbonate.

Although under most conditions the organic matter is completely lost, in the oxygen-poor conditions that formed the black shales at the bottom of the Pennsylvanian and Cretaceous seas, the muscles, organs, and skin of fish carcasses that sank into the mud did not decompose normally. They are preserved in these shales as thin films of carbon that sometimes reveal the outline of the entire body. This process of carbonization is the most common way in which plants are fossilized.

In a few very rare cases, mammoths and other mammals from the Ice Age have been entirely preserved by freezing or drying out to form a natural mummy, but nothing like this has ever been found in Nebraska. The climate here has been too warm and wet to let such fossils survive.

Precambrian Time

The rocks that formed early in earth's history in Precambrian time are generally exposed in North America in two types of settings: areas that have experienced extensive continental glaciation such as Minnesota, or in areas that have undergone extensive uplift and erosion such as the Rocky Mountains. In both cases, the younger overlying sedimentary rocks have eroded off the older underlying basement rocks, exposing them.

In Nebraska, there are no exposures of Precambrian rocks. Great thicknesses of sedimentary rocks blanket them, and yet, we think we know a fair bit about their distribution, age, and origins. How do we know this when we can't even see the rocks? Geologists rely on evidence gleaned from geophysical surveys and well drilling.

People looking for geologic resources, such as oil and gas, and for geologic knowledge have drilled many wells in Nebraska. During the drilling process, the drill rig operators recover bits of rock fragments that identify which rock formation they are currently passing through. If they drill deeply enough, they will eventually penetrate the Precambrian rocks; about 19,000 wells in Nebraska reach the basement rock. By studying the mineral make-up and chemical composition of the recovered bits of rock, a geologist can determine what rock type is present. Isotopes of radioactive elements can be analyzed to determine the geologic age of the samples. It is possible to learn quite a bit about an area from drill hole samples.

In addition to direct sampling, a variety of techniques have been developed to "remotely sense" these ancient rocks. By studying small changes in the surface magnetic field and gravitational field, geologists interpret what rock types are below the surface. Changes in the mineralogy of the rocks at depth cause changes in these geophysical signatures. A buried rock body with enough magnetite can cause a distinct anomaly in the magnetic field at the surface, and a large rock body composed of denser minerals than those of the surrounding rocks produces an anomaly in the gravitational field detectable at the surface.

The oldest rocks under Nebraska are metamorphic rocks, mainly gneisses, that are similar in age and composition to rocks exposed in the Rocky Mountains of Colorado and New Mexico. We think that they are part of a broad belt of similar rocks that underlie New Mexico, Colorado, Kansas, Missouri, and Iowa. These metamorphic rocks formed in the interior of a mountain belt some 1.8 to 1.6 billion years ago. This mountain-building event, known as the Central Plains orogeny, occurred near the end of a long series of major events during which the North American continental interior was assembled from smaller pieces of continental crust. Part of one of these boundaries between smaller pieces occurs in northwestern Nebraska and is known as the Colorado-Wyoming lineament. Marv Carlson of the Nebraska Conservation and Survey

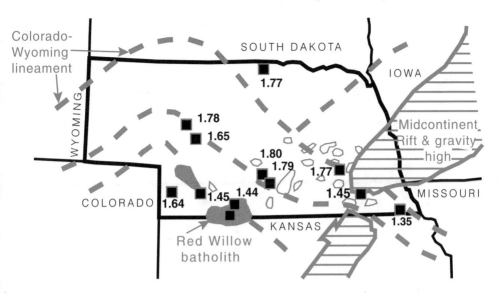

Major Precambrian basement elements in Nebraska. Boxes are drill hole localities where samples have been dated, and numbers are age in billions of years. Solid gold shapes are 1.4-billion-year-old granites that are not associated with any specific mountain-building event. Open gold shapes are magnetic anomalies that could be similar granites, but they have not been sampled or dated. —*Modified from Van Schmus and others, 1992; Carlson, 2000*

Division postulates that three such boundaries exist in Nebraska. Shortly after the Central Plains orogeny, the interior of the continent stabilized and has remained a coherent entity. However, these boundaries may have controlled minor faulting and crustal warping that occurred at much later times. For example, the Toadstool fault in northwest Nebraska that moved sometime after 30 million years ago is along the Colorado-Wyoming lineament.

Quartzite, a metamorphosed sandstone, exists in the subsurface at several localities in Nebraska. It is tempting to correlate the quartzite with similar looking and distinctive rocks—the Sioux and Baraboo quartzites—exposed in South Dakota, southern Minnesota, and Wisconsin. Unfortunately, it is impossible to determine if they are the same age because there are no fossils present and no minerals in these quartzites that can radiometrically date the time of deposition or metamorphism.

About 1.4 billion years ago, large masses of granitic magma intruded the crust in Nebraska, as well as many other areas in the midcontinental region. These granites seem to have intruded when the crust was stable; they are not associated with mountain building like many other granites. Some geologists believe they formed in a unique event in the earth's history, which hadn't happened before and hasn't happened since, and their origin is still not well understood. In Furnas County of southwestern Nebraska, a large intrusion called the Red Willow batholith was recognized from drill hole and magnetic data. Many of these granites produce large magnetic anomalies because they are particularly rich in magnetite.

The last major Precambrian event in Nebraska, a failed attempt to rift the North American continent apart, occurred about 1.1 billion years ago. This left behind a large rift valley called the Midcontinent or Keweenawan Rift. This geologic feature is exposed in the Lake Superior region of Minnesota, Wisconsin, and the Keweenaw Peninsula of Michigan, and extends to the southwest in the subsurface through western Iowa, eastern Nebraska, eastern Kansas, and perhaps into Oklahoma. A large positive gravity anomaly known as the midcontinent gravity high reveals the feature, which is otherwise undetectable at the surface. As the continent was ripping apart, the rift valley sank and subsided thousands of feet or more. Magma from deep within the earth rose through faults that edged the valley. The magma erupted as lava over the surface and solidified into basalt, a dense, dark, iron- and magnesium-rich volcanic rock. These basalts create the midcontinent gravity anomaly. Associated with the basaltic eruptions were minor eruptions of rhyolite, a light-colored silica-rich volcanic rock. Eruptions along modern continental rifts, such as the East African Rift, also produce a lot of basalt and some rhyolite.

As the rift valley sank, it filled with great thicknesses of sediments, which ultimately buried it. In other parts of the world, these types of sedimentary deposits have hosted important deposits of oil and natural gas. Shales rich in organic matter are exposed in the rift in Wisconsin. Several oil companies have intensely studied the Midcontinent Rift sediments and drilled test wells across the border in Iowa, but they have not found economically recoverable deposits.

While the features of the buried rift are not exposed at the surface, they do have an indirect surface manifestation. A series of faults, collectively referred to as the Humboldt fault zone, occurs in the southeastern part of Nebraska, above the buried rift. Faults are younger than the youngest rocks they cut, and the

Humboldt fault zone cuts rocks much younger than the rift sediments. Indeed, a concentration of earthquake activity suggests these faults may be presently active, albeit moving slowly and only producing small, infrequent earthquakes in historic times. The spatial association of younger faults and earthquakes with old rifts within a continent is a common one, and geologists think that old structural flaws deep in the basement foundation act as weaknesses that get reactivated at times because of forces within the tectonic plates. This might be considered broadly similar to how cracks can reappear in a plaster wall after it has been painted or repaired. In this way the underlying Precambrian foundation continues to influence geologic activity.

There is a large gap in Nebraska's geologic history from 1 billion years ago to the start of Paleozoic time some 530 million years ago. Erosion during this time did not cut deeply enough to erase the rift sediments, but we know very little about these 500 million years.

Paleozoic Time

At the beginning of Paleozoic time, shallow seas invaded the continent. As they swept in, they deposited beach sands directly on top of the eroded gneisses, quartzites, and granites of Precambrian time. Because of the relative stability of the midcontinent region since Precambrian time, these first beach sands, now sandstone, remain at the bottom of broadly horizontal sedimentary strata over large areas. During this long period of stability, the older rocks were progressively covered by overlying younger rocks.

The layers are not perfectly horizontal, dipping very gently to the west in eastern Nebraska. In addition, the present land surface is largely erosional and slopes to the east, dropping about 4,000 feet from the Panhandle of Nebraska to the Missouri River. Consequently, the lower, older layers in the sedimentary stack are exposed at the lower elevations in the southeastern part of the state. Even so, only the youngest Paleozoic rocks are exposed in Nebraska, and the shoreline sandstone at the bottom of the stack is known only from subsurface data.

The relative stability of the midcontinent region also means that since Precambrian time, the land surface has never been far from sea level and has had remarkably little topographic relief. Slight increases in sea level have flooded vast areas. When global

Early Permian

Chase Group (0 to 300 feet thick): variable thin limestones alternating with multicolored diverse shales

Council Grove Group (0 to 300 feet thick): multicolored shales interlayered with cherty and fossiliferous limestones, with well-developed paleosoils in southeast Nebraska

Admire Group (0 to 150 feet thick): multicolored shales alternating with very fossiliferous limestone with an upper micaceous sandstone, and overlying stromatolites

Pennsylvanian

Wabaunsee Group (0 to 300 feet thick): shale, sandstone, and limestone layers, with minor coal beds that formed in a shallow marine and coastal setting

Shawnee Group (0 to 200 feet thick): alternating marine limestones with shales, with stromatolite and other algal structures in the limestones

Douglas Group (0 to 100 feet thick): gray to black to red shales with some limestone

Lansing Group (0 to 75 feet thick): alternating cherty marine limestones and shales

Kansas City Group (0 to 200 feet thick): alternating marine limestones and shales, includes Winterset limestone, a 24-foot-thick unit often quarried for crushed rock in Omaha area; four old soil horizons mark emergent part of cyclothems

Paleozoic rocks exposed in Nebraska.

sea level was high, sediments accumulated in the shallow seas of the midcontinent. When sea level was low, the land was exposed and subject to erosion or the accumulation of terrestrial sediments. The thin, patchy, nonmarine sediments were most vulnerable to erosion and seldom remained. Erosion stripped away rocks deposited during significant intervals of geologic time. With crustal stability and a lack of significant basin formation and a low subsidence rate, geologic units are widespread but quite thin. Frequent gaps in deposition interrupt the dominantly marine sediments of Paleozoic time, but these are too small to be considered unconformities.

One exception to the widespread but thin sediments occurs at the southeastern corner of Nebraska. Because of the continuing instability of the Midcontinent Rift in the Precambrian basement rocks, this area sank more rapidly than the surrounding region under the load of Paleozoic sediments. The area, called the Forest City Basin, accumulated a greater thickness of sediments than the surrounding areas during the middle and late Paleozoic era. Sediments occasionally spread across the western boundary that separated the subsiding area and more stable crust. These sediments hardened into brittle rock and broke whenever the basin became unstable and sank. They form the Humboldt fault zone. Most of the Forest City Basin lies within Kansas.

Cambrian through Devonian Subsurface Rocks

During Cambrian and early Ordovician time, sea level gradually rose, pushing the shoreline farther into the continent's interior, progressively flooding more of it until only a few islands remained. The shallow sea did not reach Nebraska until late Cambrian time. The first sediments preserved on top of the Precambrian erosional surface are thin sands reworked on the beach. These late Cambrian sandstones are covered by shallow-sea deposits of carbonates that continued to be deposited into early Ordovician time. A similar sequence is visible at the surface in central Missouri, where the Cambrian sandstone is called the LaMotte formation and the Cambrian-Ordovician dolostones include several formations.

After an erosional gap, the sea deposited another similar sequence during middle and late Ordovician time. The beach sandstone at the base of this sequence is called the St. Peter sandstone and is a widespread formation, present in many midwestern states. A very pure sandstone, it consists of quartz grains of uniform size with little cement. Because it is porous and permeable, it makes a good reservoir for petroleum or natural gas. Oil companies have produced some oil from this sandstone and from the fractured limestone of the Viola formation—a late Ordovician carbonate on top of the sandstone—in southeastern Nebraska. The St. Peter sandstone beneath the surface near Omaha has been used as a storage "tank" for natural gas, which is pumped directly into the ground from a pipeline.

Another gap in the rock record separates the Ordovician carbonates from the remnants of Silurian dolostone, and yet another period of erosion removed some Silurian rocks before deposition of Devonian carbonates. The Devonian rocks were in turn eroded before the return of the seas during Mississippian time.

Late Paleozoic Arches and Basins

The end of Mississippian time and early Pennsylvanian time were restless for the continent, even for the midcontinent interior and Nebraska. To the east, a tectonic plate collision was forming the Appalachian Mountains. To the south, the Ouachita Mountains were rising in Arkansas and Oklahoma, and out west the Ancestral Rockies orogeny was building mountains. Although we have been describing locations in terms of present-day geography and compass directions, this reference system was entirely different during Mississippian and Pennsylvanian time. The Appalachians rose when the North American continent was near the equator, partly submerged, and surrounded by tropical seas.

While no mountains formed in the interior of this tectonic plate, forces associated with these events were strong enough to reactivate faults and warp the land into large arches and basins. The Cambridge Arch, a subsurface feature that extends diagonally across western and central Nebraska into Kansas, divides areas where thicker sequences of sediments occur. The Humboldt fault zone, which is exposed in the southeastern corner of the state, was particularly active. With the Nemaha Uplift rising on its west side and the Forest City Basin sinking on the east, the fault moved almost 3,000 feet at this time. Erosion removed sediment from the tops of these arches and deposited it in the basins.

Oscillating Seas of
Pennsylvanian and Permian Time

During middle to late Pennsylvanian time stability was restored to the continent's interior and shallow seas deposited marine sediments once again across Nebraska. Because of the tectonic stability,

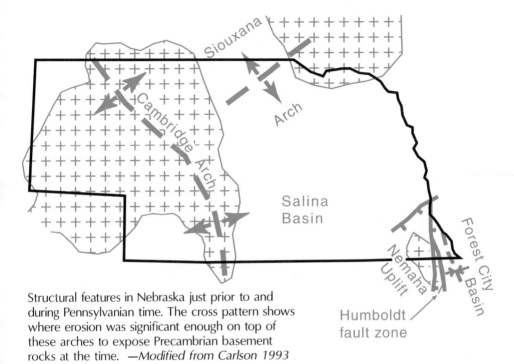

Structural features in Nebraska just prior to and during Pennsylvanian time. The cross pattern shows where erosion was significant enough on top of these arches to expose Precambrian basement rocks at the time. —*Modified from Carlson 1993*

distinctive rock layers are often quite thin, sometimes a foot thick or less. But, because of the extremely low relief of the surface, similar conditions persisted over large areas, and these thin layers can be traced for long distances, even for tens of miles. The sea receded from time to time, exposing the marine sediments to the atmosphere, as indicated by ancient soils and some nonmarine layers, including coal.

Limestone and shale make up the great majority of the Pennsylvanian rocks but each comes in a variety of types that indicate different sedimentary environments. Some limestones are dense, fine-grained, and only sparsely fossiliferous; others are silty and crammed with fossils. Some shales are red and blocky; others are black and thinly laminated. Through careful study of these rocks and their fossils, we have found that these differences are largely the result of varying water depth or distance from the shoreline. The various kinds of limestones and shales are not stacked in a haphazard fashion but instead tend to follow a repeated sequence along the lines of ABCB—ABCB—ABCB, where A is gray shale and terrestrial sediment of nearshore environments, B is limestone of

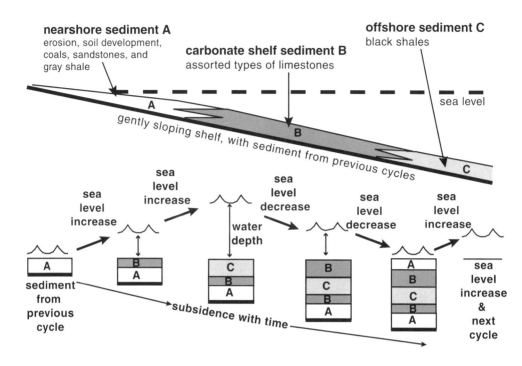

Cyclothems develop with oscillating sea level and shifting sites of deposition.

shallow sea environments, and C is black shale of deep water environments. In a vertical stratigraphic sequence the strata describe a history of repeated cycles of rising and falling sea levels, shifting shorelines, and different depositional environments. Geologists have dubbed the cycles *cyclothems*.

When sea level was low, the sea's eastern coastline was probably within or near the eastern part of Nebraska. The Appalachian Mountains had already been built by the collision of continental plates during the assembly of the supercontinent Pangaea, and rivers carried mud and some sand down from them and deposited it in deltas and marshes along the shore. In some stagnant marshes, enough plant debris accumulated to produce peat that would eventually become coal. As sea level began to rise, waves washed over these coastal environments and sediments, eventually submerging them, creating estuaries, and pushing the coastline back toward the mountains. Sea creatures invaded with the rising sea and lived in the shallow coastal waters on the muddy bottom.

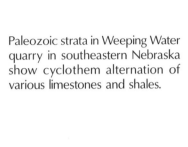

Paleozoic strata in Weeping Water quarry in southeastern Nebraska show cyclothem alternation of various limestones and shales.

These coastal environments produced the gray shales (A) of the cyclothems.

The rising sea was still shallow as seas go, but the feeble supply of clay from the now distant rivers was so slight that the water was clear, allowing sunlight to reach the bottom. Photosynthesizing algae and organisms with skeletons of calcium carbonate flourished. When they died, they sank to the bottom and broke into fragments, producing the lime mud that lithified into limestone. Sometimes storms swept across the platform and deposited coarser debris. With Nebraska in an equatorial position at that time, the tropical shallow seas were a good environment for carbonate production, which generated the carbonate layers (B) in the cyclothems.

As the sea continued to rise, the water deepened and isolated the bottom water from the surface and any opportunity to replenish its oxygen. The water didn't have to be very deep for this to happen if warm surface water created an upper layer that—though well stirred by wind and waves—never mixed with the colder water at the bottom. As the remains of dead organisms sank to the bottom and decomposed, the oxygen was used up. Eventually, there wasn't enough oxygen to decompose organic matter, and it accumulated in the colder still water along with what little fine clay had drifted far out into the sea before sinking. This organic mud turned into the distinctive black shales (C) of the cyclothems that are sandwiched between limestone layers.

Falling sea level, an influx of sediment, or a combination of the two eventually caused the sea to become shallow again. Oxygen and life returned to the bottom, and carbonates and the muds that created the gray shales were deposited to produce the upper part of the cyclothem. In places, rivers carrying sediment from the Appalachian Mountains built broad deltas out into the sea, pushing the shoreline back to the west. Falling sea level generally increases stream gradients and the size of the drainage basin, which in turn increases sediment supply and builds out the shoreline. When sea level again reached its lowest point, the former seabed was exposed to the atmosphere. Soil began to develop in the shales, and in places nonmarine sediments were deposited.

Geologists first recognized these cycles in Pennsylvanian rocks in Illinois, where about half of the sequence consists of marine rocks and about half nonmarine rocks. The nonmarine environments include rivers, deltas, and coastal lowlands that provided the wetlands that produced coals. Tracing the cyclothems from Illinois eastward, one finds that the nonmarine sediments represent an increasingly greater proportion of the cycles. Marine sediments make up the greater proportion of the sequence in rocks farther west, including those in Nebraska. The erosion of the rising mountains to the east produced sediments that poured into the sea and onto a broad continental shelf extending to what is now the Great Plains. The sediments that reached Nebraska from this far mountainous source were mostly fine mud, with only occasional sand. Nebraska was also at such a low elevation that it was submerged more often than the continental shelf to the east, so marine rocks dominate the Nebraska section of Pennsylvanian rocks.

What caused sea level to rise and fall with such seemingly regular cycles? Although geologists have long debated this question, they now think that growing and shrinking glaciers of a south polar ice cap were responsible. As the glaciers expanded, holding more and more water as ice, the level of the world's oceans fell. During ice cap melting, the oceans overflowed, flooding more of the continent and broadening the continental shelves. Abundant evidence from rocks of that geologic time in southern Africa, South America, Australia, Antarctica, and India indicates a south polar ice cap covered part of the supercontinent Pangaea that these continents were part of.

While the growth and shrinkage of a polar ice cap explains falling and rising sea levels, what caused the ice cap to expand and retreat? The ice ages of more recent earth history show that climatic changes were responsible for the glacial cycles of advance and retreat. Astronomical cycles involved in the relationship between the earth and the sun appear to play a role in climatic changes. Named Milankovitch cycles after the climatologist who first proposed an astronomical explanation for glacial advances and retreats, these cycles affected solar input and thus earth's climate, causing sea level fluctuations, and ultimately, cyclothems.

In the subsurface of the far western part of Nebraska are nonmarine redbeds, sediments stained red by small amounts of iron oxidized in the atmosphere. They formed around crustal welts known as the Ancestral Rockies that remained above sea level in the area that is now Colorado. In the shallow seas around the Ancestral Rockies, gypsum and salt deposits accumulated where salt flats formed in an arid climate.

In many ways, the Permian rocks of Nebraska follow the pattern of the Pennsylvanian rocks. Cycles of different sedimentation in varying water depths continued, but sea level was generally lower during this period of time and covered less of the continent. The northward drift of Pangaea throughout late Paleozoic time carried North America into the Northern Hemisphere and through the tropics to the latitude of subtropical deserts. There was more dry, arid land. The Permian rocks of eastern Nebraska consist of shale, limestone, and sandstone—for example, the Indian Cave sandstone exposed at Indian Cave State Park along the Missouri

River. Coals are absent and the soils that formed on sediments that were exposed above sea level indicate an arid climate. The Permian rocks in the subsurface farther west include gypsum and salt, evaporite products of drying seas. The gradual shift to more terrestrial conditions continued, and in eastern Nebraska, erosion slightly outpaced deposition, producing a gap in the stratigraphic record.

Late Paleozoic Life

The Paleozoic era is named for "ancient life" because the fossils of this time interval are dissimilar to present-day organisms. In particular, most groups of marine invertebrates that are common in Paleozoic rocks are very different from any organisms of the modern world. Some of this unfamiliarity is the result of 200 million years of evolutionary changes, but much of the difference results from the extinction of many groups at the end of Paleozoic time. The largest extinction event in earth history marks the Paleozoic-Mesozoic boundary and primarily affected organisms that dwelled in the shallow seas. Much of the sea life in Pennsylvanian and Permian time consists of the last representatives of their clans.

The limestones and shales of late Paleozoic time represent diverse environmental conditions caused by fluctuating sea levels. Distinctive groupings of fossils represent the different water depths and bottom conditions.

Shallow Marine Communities

The oxygenated bottom conditions that produced gray shales and carbonates allowed a variety of organisms to live on and in the bottom sediment. The fossils in these strata are primarily of invertebrate organisms—their carbonate skeletons were preserved in the shallow, oxygenated environments where decomposition of

In this hash of fossil fragments are two larger brachiopod shell fragments and many small rice-shaped shells of fusulinids.

organic matter was rapid. Vertebrate remains aren't as common in the Paleozoic limestones and gray shales as are the invertebrates, but you can find the teeth of sharks and their relatives.

Marine Invertebrates

Fusulinids: Many one-celled organisms construct a hard skeleton that may be preserved in the fossil record, but fusulinids are particularly distinctive. These organisms had an elongate coiled shell that resembles a tiny football, an appropriate fossil for the Cornhusker State! They appeared and flourished in late Paleozoic time, evolving rapidly and achieving large sizes—up to about an inch long—for so simple a creature. Fusulinids about the size and shape of grains of rice are common in some Pennsylvanian rocks in Nebraska, and sometimes they make up most of the rock. They were mobile but lived on the bottom muds eating other tiny organisms. In spite of their success and apparent abundance, the fusulinids went extinct at the end of Paleozoic time and have no living descendants.

Corals: The group that includes all modern reef-building corals does not show up in the fossil record until Mesozoic time, but groups that are distantly related to modern corals inhabited Nebraska's Paleozoic seas. The rugose corals include some species that formed colonies, while others remained solitary individuals. Anemone-like solitary corals sat just above the seafloor affixed to the top of a curved, conical external skeleton of their own construction, a highly preservable structure that the group is popularly named for—horn corals. The fossils reach several inches in length along their axes and are the Nebraska state invertebrate fossil. Some colonial species produced massive structures of fused skeletons; others made a snarl of cylindrical skeletons, each about a centimeter or more across, that trapped sediment in the openings between skeletons to form a massive structure, part sediment, part skeletal.

Another major group, the tabulate corals, included only species that formed colonies. The individual animals were usually smaller than the rugose corals, building tubular skeletons that were only a few millimeters across, but they fused or twisted them together to form structures comparable in size to the rugose coral colonies. Both of these groups of corals, which had been common throughout most of Paleozoic time, were extinct by the end of that era.

Individual horn coral from late Pennsylvanian time of Nebraska. When the organism was living, the tapered end of the skeleton would have been embedded in the bottom sediment with the long axis standing upright in the water and the coral polyp sitting atop the open end.

Bryozoans: Bryozoans, another group of colonial organisms, bear a superficial similarity to coral colonies, albeit on a much smaller scale. However, bryozoan animals are more complex than the coral animals and belong to an entirely different group of organisms. Individual bryozoan animals of Paleozoic time were tiny, only about a millimeter or less across, and the colonial structures they built were relatively small. Some species formed little hemispherical lumps; others constructed branching, twiglike shapes, all covered with tiny openings that sheltered the little animals. Some bryozoans created lacy curtains of calcite on which to live. Most of the common bryozoans, including those mentioned here, went extinct at the end of Paleozoic time, but some less prominent ones survived and flourished in Mesozoic time, and their ancestors flourish today.

Brachiopods: Brachiopods are a group of marine, shelled creatures that, even though they survive today, are hardly well known to the average person. Modern brachiopods are confined to marine environments seldom visited by humans. What's more, with little meat in their shells, they are not a common seafood item. However, in Paleozoic time they were prolific and displayed a remarkable variety of forms. Brachiopods are closely related to bryozoans, but superficially resemble clams. One difference between brachiopods and clams is that the lower and upper shells of a brachipod are often very different, while clams have similar shells.

Several brachiopods in Pennsylvanian limestones from eastern Nebraska. The large one to the left of the penny has been flattened. To the right of the penny, the shell of the productid, a type of brachiopod, has the broken bases of spines along the straight edge of the shell where it hinged together with its counterpart.

In addition, individual clam shells are asymmetrical—they are left handed or right handed—while individual brachiopods shells are symmetrical. Brachiopods fed by circulating water over a complex filter-feeding structure in the space between their shells.

The productids are a common brachiopod group in Pennsylvanian rocks of Nebraska. The distinctive lower shell is swollen to a shape like a teacup and sprouts spines that raised the brachiopod above the soupy mud of the bottom. The spines usually are not preserved intact, but you can often see their bases. The flat upper shell served as a lid. Productids come in a wide range of sizes up to almost the size of a tennis ball, and while successful during late Paleozoic time, they were extinct by the end of the era.

Another common group of brachiopods is the spiriferids. Some spiriferids have wide, ridged, tapering shells that come to a point at each end of the hinge where the two shells maintain contact. Others have simple, smooth, rounded shells. All spiriferids have a complexly coiled feeding apparatus within their shells. Most

This Pennsylvanian mollusk shell resembles a modern scallop shell. Note the brachiopod shell fragments around it.

spiriferids became extinct at the end of Paleozoic time, but a few persisted until Jurassic time.

A group of small, morphologically primitive brachiopods evolved in Cambrian time and continue to live today. Called in-articulate brachiopods, these little circular or spoon-shaped fossils are often preserved in the dark gray shales that accumulated where there was just enough oxygen to support them, and some small mollusks, but little else.

Mollusks: Mollusks, a large and ancient phylum that includes clams, snails, and cephalopods, flourished in late Paleozoic time. Unlike the fusulinids, rugose and tabulate corals, and some bryozoans and brachiopods that went extinct at the end of Paleozoic time, mollusks were just beginning their long, successful evolution. Although clams and snails lived throughout Paleozoic time, brachiopods dominated the scene. In Mesozoic time, the bottom-dwelling mollusks took over, perhaps exploiting opportunities left open by all the extinct species. Another possible scenario is that the explosive evolution of mollusks forced many other marine organisms into extinction.

Cephalopods, predatory squidlike mollusks, swam in Nebraska seas. Some of their large, coiled shells resembled those of their modern relative, the chambered nautilus, and some had the more complex structure characteristic of the ammonites, a very successful group of cephalopods. Ammonites first appeared in Devonian time and evolved rapidly into many distinctive forms. Though their late Paleozoic diversity was drastically reduced during the Permian extinction event, some ammonites survived, evolving throughout Mesozoic time only to go extinct at the end of that era.

Arthropods: Arthropods are not as common in Nebraska strata as other invertebrates, but they can be found in certain layers. Trilobites, distinctive members of early Paleozoic communities, had already declined in diversity in early Paleozoic time and were completely extinct by the end of Permian time. However, fossils of the last surviving group of trilobites are preserved in the Pennsylvanian and Permian rocks of Nebraska. Much the same can be said for the eurypterids, marine relatives of scorpions, except that they were never as diverse as the trilobites.

Echinoderms: The sea urchin is a modern echinoderm. Its ancestors occur as fossils in Nebraska's Paleozoic strata. Their skeletons are seldom preserved intact and instead the plates and spines litter some layers.

By far the most common echinoderm fossils in the late Paleozoic rocks are pieces of crinoids, which resemble plants; modern crinoids are called sea lilies. Each crinoid is attached to the sea bottom by a stalk or a stem. At the other end of the stem, its compact body is surrounded by an array of feathery arms that spread to capture sinking or drifting food particles in the water. This entire structure is supported by an internal skeleton of calcium carbonate that consists of numerous calcite plates. The greatest number of these plates are similarly shaped perforated discs that are stacked on top of one another to make up the stem.

When a crinoid dies and decomposes, the skeletal plates break free and are buried separately. Crinoid plates are common fossils in late Paleozoic limestones. In fact, some limestones are composed largely of crinoid fragments. Most crinoid groups went extinct

Crinoid plates and stalk sections from Nebraska's Pennsylvanian strata.

at the end of Paleozoic time, and although one evolutionary branch thrived and is still successful today, crinoids never again dominated shallow marine environments.

Fossils from the Black Shales

The black shales in the Paleozoic rocks were deposited in the deep parts of the sea. The bottom had so little oxygen that virtually nothing lived there, but that doesn't mean there aren't any fossils. When creatures swimming in the waters above died and sank to the bottom, their carcasses remained unmolested by scavengers. Even soft parts like cartilaginous skeleton, muscle, or skin, were often preserved as a thin film of carbon in the shale. The black shales give us the best record of some organisms whose remains were often destroyed in oxygen-rich environments, where decay is rapid and scavengers are efficient.

Invertebrates

Crustaceans: A crustacean is a type of arthropod. Arthropod remains were not readily preserved in the shallow environments, with the exception of sturdy trilobite skeletons. But the black shales preserved shrimplike crustaceans intact. Although they are invariably flattened by the compaction that shales undergo, the segments of the skeletons are usually discernible.

Vertebrates

Conodonts: The black shales are a good place to find conodonts, which are often barely visible, shiny, toothlike elements, about a millimeter or less in length. They are composed of calcium phosphate, come in a variety of shapes that bear denticles or cusps, and may be found singly or in clusters. The distinctive morphologies of conodonts changed rapidly through time, allowing geologists to use the fossils to pinpoint the approximate age of the enclosing shales. Although present in the other kinds of marine sediments of Nebraska, they are difficult to find and recover in rocks other than black shales.

The identification of these elegant little structures puzzled paleontologists for most of the twentieth century. Based on more complete specimens, paleontologists believe they are the mouth parts of small wormlike organisms. The superficial resemblance of the animal to a worm tells us little, but it appears it may be a very early offshoot of the vertebrates.

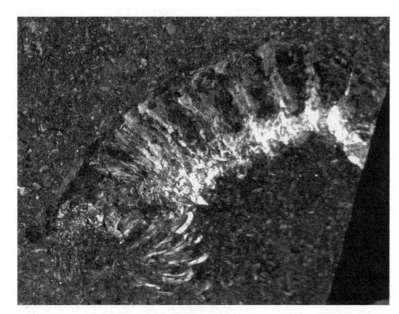

Posterior part of a Pennsylvanian shrimplike crustacean in black shale.

Sharks and Cartilaginous Fish: You can find the teeth of sharks and their relatives in limestones, but the black shales yield not only teeth but once in a while the whole shark. Patient fossil collectors split the shales along bedding planes and have revealed spines, fins, portions of carcasses, and sometimes entire fish. They are often flattened and are reduced to thin films of carbon but yield a lot of information. The black shales also preserve cream-colored carbonate nodules up to several inches long with scales, bones, and teeth scattered inside. These are coprolites, otherwise known as fossil feces, and are probably the remains of some shark's lunch.

Sharks and other cartilaginous fish are perhaps the most common vertebrate fossils in the black shales. They range from just an inch up to several feet in length. Some had teeth with a broad, flat crown used to crush shellfish; others had the three-pointed cusps of a primitive shark. One unusual group of sharklike fish, the iniopterygians, had elaborate spines and specialized organs called claspers that were probably used in mating, although it is difficult to say exactly how. If we had not discovered such completely preserved fossils of these fish, it is doubtful that we could ever have imagined that such creatures existed.

Bony Fish: Black shales also preserve fossils of fish that had bony rather than cartilaginous skeletons. These ancient relatives of modern ray-finned fish were covered by thick, bony scales like those of the present-day gar but they exhibited a range of body forms similar to that of more advanced fish, from the torpedo shape of a trout to the deep, narrow body of an angelfish.

Freshwater and Terrestrial Communities

When sea level was low during Pennsylvanian and Permian time, some sediments were deposited in freshwater and estuarine environments preserving fossils of freshwater organisms. Based on an association with sediments of freshwater environments, paleontologists believe that xenacanth sharks, an extinct group, were freshwater fish. They had an unusual body form for a shark, with very unsharklike tail and fins, and teeth with two unmistakable pointed cusps of about equal size. Lungfish and amphibian fossils have also been found in the rocks of Permian time.

Shark tooth in Pennsylvanian black shale. The tooth size and flat crown tells us it was probably used for crushing shellfish.

The fossil plants from the Pennsylvanian and Permian rocks of Nebraska include giant horsetails, ferns, and tree ferns. Most of the plants that are familiar to us today were just beginning to evolve or would not appear until much later in the history of life.

Cretaceous Time

In eastern Nebraska brown and red sandstones of the Dakota group overlie the Pennsylvanian limestones and shales in places. The contact between them is exposed in a number of bedrock quarries and outcrops along the lower section of the Platte and Elkhorn Rivers, such as at Schramm Park. The sandstone is of Cretaceous age, strata from Triassic and Jurassic time of the Mesozoic era are missing. This stratigraphic gap of some 140 million years is a major disconformity over the eastern half of Nebraska, reflecting a time when the land was above sea level and erosional forces prevailed.

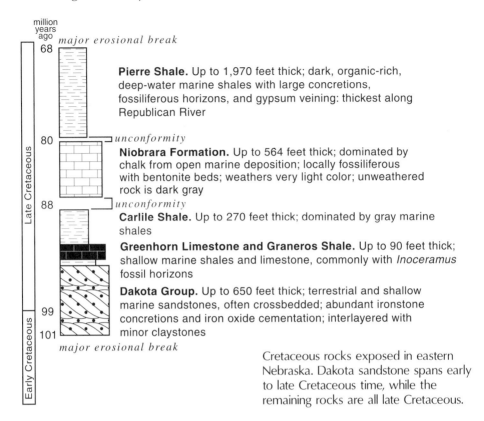

million years ago

major erosional break

68

Pierre Shale. Up to 1,970 feet thick; dark, organic-rich, deep-water marine shales with large concretions, fossiliferous horizons, and gypsum veining: thickest along Republican River

80   *unconformity*

Niobrara Formation. Up to 564 feet thick; dominated by chalk from open marine deposition; locally fossiliferous with bentonite beds; weathers very light color; unweathered rock is dark gray

88   *unconformity*

Carlile Shale. Up to 270 feet thick; dominated by gray marine shales

Greenhorn Limestone and Graneros Shale. Up to 90 feet thick; shallow marine shales and limestone, commonly with *Inoceramus* fossil horizons

Dakota Group. Up to 650 feet thick; terrestrial and shallow marine sandstones, often crossbedded; abundant ironstone concretions and iron oxide cementation; interlayered with minor claystones

99

101

major erosional break

Cretaceous rocks exposed in eastern Nebraska. Dakota sandstone spans early to late Cretaceous time, while the remaining rocks are all late Cretaceous.

Triassic and Jurassic strata exist in the subsurface in western Nebraska, and farther west much more complete sections are exposed in the tilted strata along the flanks of the Rocky Mountains. The Dakota sandstone is at the base of a varied Cretaceous section that spanned some 40 million years, before another period of erosion resulted in another major disconformity, this time throughout Nebraska.

The Cretaceous strata in North America indicate the interior of the continent was covered by marine waters from the Gulf of Mexico up to the Arctic Ocean, with eastern shorelines in Iowa and western shorelines deep in Colorado. This ancient water body is known as the Western Interior Seaway. The Dakota sandstone of eastern Nebraska is part of the eastern coastal complex of sediments that formed as the seaway first spilled into the area and migrated east.

Dakota sandstone in quarry wall along the lower Platte River. Note the layers that are inclined downward to the left (west)—these crossbeds indicate that the current flowed from east to west during deposition. At least five different layers, with crossbeds inclined in the same direction, are stacked here. Darker layers have more iron cement than lighter layers.

The Paleozoic shales and limestones beneath the Dakota sandstone do not look normal for tens of feet deep, especially the shales, which are red and green instead of the usual black or gray. This color change is part of a buried weathering profile, an old soil from Cretaceous and older times. Such soils provide information on the climate then, which in this case was subtropical.

Cretaceous Stratigraphic Units

Cretaceous deposits are present in all but the southeastern corner of Nebraska. They thicken towards the west, where they are more than 2,000 feet thick, as revealed by oil wells. Over most of their extent in Nebraska, the Cretaceous strata are buried under Cenozoic rocks and sediments.

Dakota Group: The Dakota group, informally known as the Dakota sandstone, is dominated by a variety of brownish red sandstones, with some thin conglomerate lenses and local claystone lenses. The different degrees of cementation in the sandstone affect the color. Where there is little or no cement, the sandstone is lighter in color and can be broken apart by hand. Water easily moves through this poorly cemented and porous sandstone, creating a useful aquifer and springs. In western Nebraska, beneath the surface some 4,000 to 6,000 feet deep, the Dakota sandstone contains oil and gas.

In some distinct horizons within the sandstone, iron oxides are particularly densely concentrated to form dark reddish brown ironstone layers. The iron oxide not only cements the sand grains, but partially replaces them. It also grows as nodules and other irregular forms within the layers. Some of these ironstone concretions can be mistaken for fossils. In a few localities, iron oxides replaced the woody tissue of logs, preserving them as fossils. The ironstone horizons probably formed when soil developed in the recently deposited sands. Such cementation associated with soil development occurs in modern tropical soils known as oxisols. Iron oxides can also form at the top of a shallow groundwater table.

Conglomerate layers within the sandstone contain abundant weathered chert clasts that came from chert nodules that were in the underlying Paleozoic limestones. These fragments were eroded from exposures to the east and north and transported to their final resting place by rivers and shoreline currents. Crossbeds, inclined layers that form when sand is deposited by a current, are common in the Dakota sandstone, as are channels that were scoured by water currents and then filled with sediment. Orientations of the crossbeds and channels indicate which way the currents flowed. The Dakota sands were deposited on a broad, sandy coastal plain and along a high-energy coastline of the Western Interior Seaway.

Greenhorn Limestone, Graneros Shale, and Carlile Shale: The Greenhorn limestone was deposited during the first cycle of seaway development in eastern Nebraska. As the sea expanded over the continent's interior, limestones and shales were deposited on top of the shoreline sands. The depositional environment shifted continuously, forming relatively thin layers of interbedded limestone, shale, and sand. Good exposures occur at Ponca State Park.

Niobrara Chalk: The Niobrara chalk sits between two disconformities. It is best exposed in northeast Nebraska as cliffs along the

Missouri River in Cedar and Knox Counties but also occurs in outcrops in south-central Nebraska downstream of Harlan County Lake. This chalk is a distinctive deposit made of tiny platelets called coccoliths that are so small you cannot see them with an ordinary microscope. The platelets are from a planktonic algae that prospered at this time. Similar chalks make up the white cliffs of Dover, England. Widespread algal blooms were common in the warm, extensive, shallow seas of Cretaceous time and resulted in chalk deposits around the world. The bones and shells of larger animals that once swam these waters are well preserved because the entombing chalks provide an alkaline chemical environment conducive to their preservation, and because the fine-grained material preserves detailed impressions. Some of the fossils are on display at the University of Nebraska State Museum in Lincoln.

Pierre Shale: One of the best-known Cretaceous formations in North America is the Pierre shale. In Nebraska, this widespread shale is well exposed along the lower reaches of the Niobrara River valley, and in the extreme northwest corner, but it underlies deposits of Tertiary age throughout much of the state. The Pierre shale extends well up into South Dakota, encircling the Black Hills, and also flanks the Front Range of Colorado. The dominant rock type is a black shale, with some very fossil-rich and concretion-rich horizons, that was deposited 80 to 70 million years ago.

The black shales, although not striking to look at, tell us something interesting about the environmental conditions at the time of deposition. High organic content gives the shales their color. Organic material is normally recycled by bacteria and other organisms if oxygen is available. In oxygen-poor waters, the rate of decay and recycling is slower and organic material can accumulate in the sediment. However, the water above must have had plenty of oxygen to support the life that provided the organics to the bottom. In addition, the Pierre shale preserves fossils of organisms, such as ammonites, that required well-oxygenated waters. With this information, we know that the shales were deposited as the muddy, oxygen-poor, fetid bottom of a large inland sea that had abundant life in the waters nearer the surface.

Certain horizons in the shale are rich in fossils. Occasionally, the poisonous bottom waters—poor in oxygen and often rich in hydrogen sulfide—were stirred up and reached the surface. In the modern world this kind of overturning kills fish, and it likely did so in the past. The carcasses settled to the bottom and formed a fossil-rich horizon.

Black shales are a common source rock for oil and gas, which are formed by geologically "cooking" the rock. When these organic-rich sediments are heated to the right temperature range, the organic matter breaks down into oil or gas. As it migrates through the rock, some of it gets trapped in overlying or adjacent rocks. Not surprisingly, oil companies are very interested in all aspects of black shales. Where exposed, Nebraska's Pierre shale has not been buried deeply enough to cook oil and gas out of it.

The Pierre shale consists of very fine-grained sediment with a significant amount of volcanic ash that settled out of still water. Plumes of volcanic ash originated somewhere to the west of the Western Interior Seaway. Layers rich in ash are light colored and contain bentonite clays. Much of the volcanic ash, which is unstable, weathers with time in the presence of water and forms these clays.

Other distinctive geologic features within the black shales are concretions. These harder areas of rock, enclosed within the shales, come in a variety of shapes: sometimes they are perfectly round and golf-ball size, but more commonly they resemble round pillows. They can measure more than 6 feet in diameter. One distinctive type, known to geologists as septarian concretions and to collectors as turtle or thunder egg rocks, contains an interior patterned with veins of multicolored calcite bands. When broken or sawed in half, the patterns are striking.

Concretions develop when ions in the pore water of the accumulated sediment react chemically with a "seed," resulting in localized mineral growth. Sometimes fossils are found in the core of the concretions. The organic matter or skeletal material of a dead organism can form a local chemical environment that is conducive to the growth of siderite—an iron-rich carbonate. The mineral growth eventually encloses the fossil. Septarian concretions form when the concretions crack and calcite crystals grow in the cracks. You can find septarian concretions in the Pierre shale in northwestern Nebraska.

The Pierre shale produces a unique signature on the land surface—slumps and slides. The clay in the shale, especially that derived from volcanic ash, produces a very weak rock. Loose sand can maintain a slope close to 30 degrees, but the Pierre shale can't support a slope much over 10 degrees. Large slumps of overlying sediments, sliding on a base of Pierre shale, are common along the Niobrara River where it has channeled down into the Pierre shale. These slumps commonly have an arcuate or bowl-shaped upper depression and an irregular toe. This slumping is a great source of frustration for civil engineers.

Western Interior Seaway

Cretaceous strata in Nebraska are important for understanding the history of the eastern margin of the Western Interior Seaway. You might be tempted to think of the sequence, from the shoreline sands of the Dakota group to the deep water deposits of the Pierre shale, as one continuously growing and deepening seaway. However, the disconformities and detailed stratigraphy indicate otherwise. Geologists think that the sea significantly expanded and shrank some five to eight times. The Dakota sandstone and Greenhorn limestone represent the first cycle of seaway development in eastern Nebraska. The Niobrara formation with disconformities above and below, represents another cycle. Shallow seas that flooded continental interiors were common worldwide in Cretaceous time; sea level must have been significantly higher than it is now.

While these cycles may remind you of the oscillating sea level in Paleozoic times as recorded in the cyclothems, there are some important differences. Geologists have abundant evidence that in Cretaceous times the global climate was much warmer and volcanic activity was at a high level worldwide, both along submarine mountain chains known as mid-oceanic ridges where oceanic crust is created, and at large igneous provinces in oceanic and continental settings. It appears that the earth was getting rid of heat from its interior faster than usual. To accommodate this outpouring of energy, the mid-oceanic ridge system got longer and new oceanic crust spread away from the ridge faster. The older, colder oceanic crust was preferentially recycled back into the interior by a process known as subduction. On average the oceanic crust in the ocean basins became younger and hotter. The

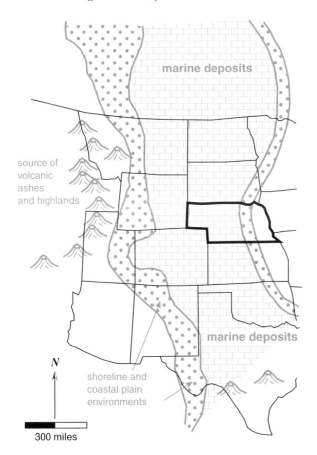

marine deposits

source of
volcanic
ashes
and highlands

marine deposits

N

shoreline and
coastal plain
environments

300 miles

Western Interior Seaway
at 83 to 79 million years
ago, spanning the time
of the disconformity
separating the Niobrara
chalk from the Pierre
shale and the beginning
of deposition of the
shale. During each Late
Cretaceous cycle the sea
expanded to cover most
if not all of Nebraska.
—*Simplified from U.S.
Geological Survey website*

hotter the oceanic crust is the higher it is, and hence, on average the ocean basins became less deep, decreasing the capacity of the world's oceanic basins to hold water. The younger, hotter oceanic crust caused the ocean waters to overflow onto the low areas of the continents somewhat like a person getting into a full bathtub, and thus shallow seas like the Western Interior Seaway were common globally at this time.

The continuous release of greenhouse gases by the unusually prolific volcanic activity may have played a role in changing the global climate. Some refer to the Cretaceous as a time when the earth was in a super greenhouse condition. We have renewed interest in what causes climate changes due to the modern global warming. The story is not fully understood yet, but we believe the greatly increased level of Cretaceous volcanism was due to the deep internal dynamics of the planet. These same deep processes may have caused anomalous activity of the earth's magnetic field during this time.

Warm shallow seas are likely to become stratified, with colder deep water isolated from the atmosphere by warm surface water. The oxygen-poor bottom conditions result in the deposition of black shales, such as the Pierre shale. Some geologists estimate that about half the earth's oil and gas formed from organic matter deposited during this relatively short interval. Interestingly, both the black shales and white chalks represent significant masses of carbon that were extracted from atmosphere and hydrosphere by photosynthesizing plankton, spent a short period of time as living organisms in the biosphere, and then were locked away in a geologic reservoir. In other words, this organic activity and sediment deposition was taking carbon dioxide out of the atmosphere, where it could cause warming, and locking it away. The net effect is to cool the climate. In this way life processes may have prevented the earth from getting even hotter. The Western Interior Seaway may have served to stabilize climate on the earth.

Life in the Seas and along the Coasts

While the Tertiary mammal fossils probably get the best press in Nebraska, the Cretaceous rocks have a rich fauna. Large clam shells, known as *Inoceramus,* are particularly common. Up to a foot or more long, these clams lived in large clusters in shallow, oxygen-rich marine waters. They are easily found at a number of sites within Nebraska such as Ponca State Park.

Ammonites, related to the modern-day chambered nautilus, were carnivorous, squidlike organisms that lived in a coiled, chambered shell and propelled themselves through the sea with jets of water. Commonly preserved in the Pierre black shales, ammonites are a group of fossils that provide geologists important information about the age of the sediments. The outside of the shells are often decorated by ridges and knobs, but often the outer layers of the shell are broken away, revealing the astounding complexity of the partitions between chambers. Some shells show rows of puncture marks that may be the bite of a larger predator, the mosasaur.

Mosasaurs, near or at the top of the marine food chain during Cretaceous time, were true marine lizards, giant monitor lizards

Inoceramus fossils from a slab of Greenhorn limestone in Ponca State Park. Sprig of raspberry for scale.

whose legs and feet had evolved into flippers. They also had a long, powerful tail for propulsion and an impressive array of teeth. Different species varied from 12 to 30 feet long. There were still larger critters, including the marine reptile *Elasmosaurus,* a plesiosaur that was 40 feet long with half of that its long neck. There were sea turtles 13 feet long and 20-foot sharks that may have weighed 20 tons. Giant fish up to 15 feet long were similar in appearance and adaptation to modern tarpon and swordfish. A fossil of a 12-foot-long fish *(Xiphactinus),* with a partially digested 6-foot-long fish in its gut, is on display in Morrill Hall at the University of Nebraska at Lincoln. Perhaps it died of indigestion.

The Dakota sandstone preserves some of the world's earliest fossils of flowering plants. Fossil tree trunks are common; iron oxide from water in the sand encrusted buried logs before the wood rotted away. Leaf fossils are preserved in clay lenses within the sandstone. Acidic soil conditions at the time of deposition of the sandstone destroyed most bones rather than fossilizing them, but a few finds, including some recently discovered tracks, assure us that dinosaurs roamed these shores. The marine rocks of late

Cretaceous time have provided evidence that pterosaurs such as *Pteranodon* circled in the skies overhead. Many of these life forms met a swift end in the mass extinction event at the end of Cretaceous time. In Nebraska, there is no record of this extinction event because erosion has removed the rocks of that age. By the time sediment deposition resumed, conditions and life forms in Nebraska had changed dramatically.

Tertiary Time

Rocks and sediments of the Cenozoic era cover most of the state and dominate the visible geology. The rocks you can see along the roads are very young in geologic terms. The Cenozoic era is divided into the Tertiary period, from 65 to 1.8 million years ago, and the Quaternary period, from 1.8 million years ago to present. The beginning of the most recent global ice ages roughly marks the division between these two periods. This is also a convenient division in the geologic history of Nebraska because the large-scale dynamic processes that transported and deposited sediment throughout the region changed at about that time.

Nebraska has been deep within the interior of the continent since the formation of the Precambrian metamorphic and igneous rocks of the buried basement. Sheltered from the spectacular tectonic processes that occurred along the western North American plate margin in Mesozoic and Cenozoic time, Nebraska has had no mountains rise up nor volcanoes erupt in more than a billion years. This lack of local tectonics means that the Cenozoic rocks lay nearly flat and remain relatively undisturbed by deforming stress. However, this doesn't mean that tectonics have not affected them. The sediments themselves provide evidence of events occurring to the west.

Seventy-five to fifty million years ago, the Laramide orogeny produced the Laramie Mountains of Wyoming, the Front Range of Colorado, the Black Hills of South Dakota, and other uplifts that form the present-day Rocky Mountains. Even as these mountains rose tectonically, erosion worked to tear them down and continued to do so long after they had finished rising. By about 10 million years ago, the debris of their erosion had buried all but the very tips of the remaining mountains and spilled eastward across Nebraska. This apron of debris, a megawedge of sediments from the Rocky Mountains, extends across not only Nebraska, but also

Kansas and Oklahoma, and down into Texas. It is more than 1,000 feet thick in places in Nebraska.

Volcanoes in the tectonically active west also contributed to the sediment in Nebraska. Volcanic ash, carried by the wind, blanketed the plains. Not only are distinct ash beds common, but much of the finer-grained sediment across the plains contains a high percentage of volcanic ash that has been reworked by wind and water. Wind-blown material, of volcanic origin and from other sources, comprises a significant percentage of the Tertiary strata.

From the beginning of the Laramide orogeny in late Cretaceous time to early Tertiary time, erosion dominated Nebraska. No rocks from Paleocene time, the first epoch of the Tertiary period, are preserved in Nebraska. The late Cretaceous Pierre shale that underlies the Tertiary sediments has been eroded and often deeply weathered into a thick soil. The earliest record of Tertiary time in Nebraska consists of a thin layer of coarse residue at the top of the Cretaceous marine Pierre shale that must have been exposed to weathering at that time. A few sparse fossils from a site in the northeastern part of Nebraska identify the age of these residual deposits as early Eocene. Only later did substantial, widespread deposits accumulate.

Although most of the Tertiary strata lie flat and undisturbed, faults cut the strata at Toadstool Geologic Park, Agate Fossil Beds National Monument, and several other areas. The distance that rocks moved along the faults varies from just a few feet to hundreds of feet. Some geologists have argued that minor tectonism continues today. Rare, small earthquakes have occurred in western Nebraska.

The faults run in two directions, almost ninety degrees in relation to each other, to the northeast and northwest. Toadstool Geologic Park is a particularly good area in which to see faults oriented in both directions. These structures probably represent minor crustal adjustments in the continental interior, a result of forces caused by changing plate boundaries. Similar faults occur in other parts of the continental interior, and some are seismically active today.

Tertiary Sediments

Geologists recognize three major pulses of deposition during Tertiary time that are represented by three major bodies of rock—

the White River, Arikaree, and Ogallala groups. Each group extended progressively farther east and south from the Rocky Mountains. The depositional agents at work were primarily rivers draining the mountains and wind blowing from the west. These processes alternately carved a system of valleys into the soft sediments and filled them again with new deposits. This cutting and filling throughout late Tertiary time superimposed one channel upon another, complicating the geology despite the lack of tectonic deformation. The simple notion of sedimentary rocks as flat uniform strata, commonly called layer-cake stratigraphy, is completely inadequate to unravel the detailed history of these sediments.

White River Group: The earliest episode of widespread deposition in Tertiary time produced the White River group of sediments. You can find this group at the surface in ravines where younger sediments have been stripped away or in the heavily eroded badlands of the high plains of western Nebraska. The lower slopes of Scotts Bluff are White River sediments, and extensive sections are exposed in the northwestern corner of the Panhandle, especially at Toadstool Geologic Park in the northwest corner of Nebraska.

The formations of the White River group span the Eocene-Oligocene boundary. The Chamberlain Pass and Chadron formations were deposited during late Eocene time, and the Brule formation was deposited during Oligocene time. The formations consist of easily eroded, fine silts and clays, with gravels and sands of ancient river channels that form ledges in steep slopes. Some less obvious channels are filled with finer muds and silts.

In the northwestern part of Nebraska, coarse gravel marks the Chamberlain Pass formation at the base of the White River group. A vigorous river system must have transported the well-rounded pebbles of igneous and metamorphic rocks from the Black Hills or the Laramie Mountains. This basal conglomerate and sand is host to a rich uranium deposit below the surface near Crawford. Groundwater flowing through the conglomerate precipitated uranium minerals into the spaces between grains. You can see these gravels exposed atop low hills of the Pierre shale several miles to the northeast of Toadstool Geologic Park.

The rivers that carried the gravel down from the mountains in the north and west carved broad valleys in Nebraska during early

Tertiary time and continued to cut valleys and fill them with the White River group sediments. The fine sediments accumulated in the wetlands of the floodplain and the sands and gravels filled the channels. Fine sediments also accumulated as windblown dust, called loess deposits. Wind also contributed volcanic ash; glass shards may make up half or an even greater proportion of the material in the finer-grained sediments. Occasionally, ash input was so rapid and abundant that relatively pure ash layers formed, and they are recognizable from afar by their distinctive white color.

Throughout the Chadron and lower Brule formations, vague dark bands, often red or brown, mark where ancient soils developed in the sediments during pauses in deposition. Although larger plant fossils are sparse, there are microscopic phytoliths, siliceous structures produced by grasses. The greenish gray sediments and soils of the Chadron formation suggest the valleys of late Eocene time were wet and forested. However, the soils of Oligocene time are red and brown and the Brule sediments are pink and tan, hues that suggest oxidation. Oxygen penetrates soils and sediments more deeply in dry conditions, so we think the climate became dryer at this time and dry woodlands among grasslands covered the landscape.

The uppermost sediments of the Brule formation consist of brown siltstones that were once considered to be part of the overlying Gering formation. They can be seen around Scotts Bluff and along Wildcat Ridge and may be traceable all the way to South Dakota.

Arikaree Group: The Arikaree group, which overlies the White River group, is exposed throughout the Panhandle. The upper part of Scotts Bluff, Wildcat Ridge to the south, and the scenic escarpment known as Pine Ridge in the northwestern part of Nebraska are part of the Arikaree group. You can also see Arikaree sediments in the valley of the upper Niobrara River at Agate Fossil Beds National Monument and where tributaries of the North Platte have stripped away younger sediments. The Rosebud formation exposed in bluffs along the Niobrara River at Fort Niobrara National Wildlife Refuge is isolated from the Arikaree group and has not been included in it, although a few fossils indicate it is of comparable age. In fact, some geologists have suggested that it should be considered part of the White River group.

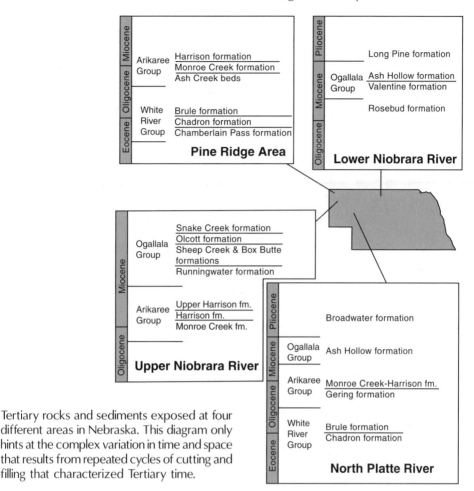

Tertiary rocks and sediments exposed at four different areas in Nebraska. This diagram only hints at the complex variation in time and space that results from repeated cycles of cutting and filling that characterized Tertiary time.

The Arikaree group was deposited from late Oligocene time into early Miocene time as volcanic ash blown into the area from the west accumulated as eolian deposits or was washed into streams and deposited by water. It includes the Gering, Monroe Creek, Harrison, and Upper Harrison formations and some informally named units. The Gering and Monroe Creek formations form the top of Scotts Bluff. A similar sequence occurs to the north, along the Pine Ridge escarpment, but there the Ash Creek beds— sediments different from the Gering—underlie the Monroe Creek formation. The Harrison and Upper Harrison formations, exposed at Agate Fossil Beds National Monument, form the top of the sequence in most places.

Snarls of thin limy structures—former plant roots and insect burrows—mark where soils developed when the accumulation of sediment slowed. In the Harrison formation, occasionally one finds a spectacular spiral-shaped burrow known as *daemonhelix* or "devil's corkscrews." Rodents *(Paleocastor)* dug these burrows, which are about 6 inches across and up to 10 feet deep, in tight, vertical spirals. This peculiar burrow form may have promoted ventilation. The sediments and fossils indicate that when the Arikaree was deposited, western Nebraska was a semiarid region crossed by small intermittent streams. Only a relatively dry climate would have permitted such deep burrows; a wetter environment would have increased the chances of flooding.

Ogallala Group: The Ogallala group, the last sheet of sediments to spread over the plains at the end of Tertiary time, consists of several formations that filled valleys initially cut into earlier Tertiary sediments by streams flowing from the Rocky Mountains. Ogallala sediments extend far into eastern Nebraska and across the entire Great Plains as far south as Texas. The sediments are mostly sands and some gravels, punctuated by infrequent volcanic ash beds— the ash originated from eruptions far to the west. The amount of volcanic material is significantly less than in the underlying Tertiary strata and may represent the waning of Basin and Range volcanic activity in the southwestern United States. Though the amount of volcanic ash in the Tertiary sediments decreased over time, major eruptions clearly affected Nebraska during late Tertiary time and produced the ash at the Ashfall Fossil Beds State Historical Park east of O'Neill. Elsewhere, the relative lack of ash and a predominance of sand and some gravel makes the Ogallala group relatively porous and permeable. These sediments hold a large, extensive reservoir of groundwater known as the High Plains aquifer, a critical source of irrigation water in Nebraska and several other states. Springs and seeps often emerge at the base of the Ogallala sediment where it is exposed. Much of the water in the Loup, Dismal, and other rivers that cross the Ogallala group strata comes from the High Plains aquifer.

Ogallala sediments, thick in places, underlie a significant part of the much younger Sand Hills. Exposures occur where the North Platte and Niobrara Rivers have cut through the sand dunes.

Devil's corkscrew, the spiral burrow of the Miocene rodent, *Paleocastor*. Note small vertical, calcified rootlets. Small dark holes are modern insect burrows.

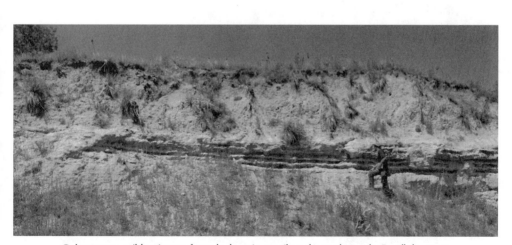

Calcareous soil horizons form ledges in a railroad cut through Ogallala strata near McCook. The uppermost of these four closely stacked, buried soils has the most calcification and thickest development.

The Ash Hollow formation, the uppermost of the Ogallala group, is one of the most widely recognized. A distinctive feature of the Ash Hollow formation is the occurrence of resistent layers called mortar beds. More resistant to erosion than other layers because of silica cement, these beds form ledges and cap rocks for buttes.

Another type of resistant ledge-former occurs where the sediments have been bound by calcium carbonate cement that forms in soils in relatively arid climates during times when sediments were not accumulating. They represent surfaces of stability rather than layers of deposition. These ancient soil horizons preserve root structures and a variety of burrows.

By the end of Miocene time, the wedge of Tertiary sediments extended from the eroded and buried Rocky Mountains like a broad gentle ramp. The top of this ramp, in many places the top of the Ash Hollow formation, represents a distinctive geologic surface. In Pliocene time, the last epoch of the Tertiary period, streams draining the Rocky Mountains once again began to cut into this wedge of sediments because of broad regional uplift, climate change, or both. Erosion exhumed the Rockies from their own detritus and stripped away most of the Tertiary sediment immediately east of the mountains. We see the mountains as they are today, rising abruptly from the plains, because of this Pliocene erosion. Thus, the Rockies have, in a sense, been reborn as topographic mountains.

In the Panhandle of Nebraska, between the north and south branches of the Platte River, a tableland preserves a remnant of the once-broad ramp of Tertiary sediments with the distinctive surface of Ash Hollow formation overlain by loess deposits. The formation type locality is at Ash Hollow State Historical Park. The surface gradually ascends to the top of the Laramie Mountains in Wyoming much as it did at the end of Miocene time. This narrow strip of sediment is called the Gangplank because it affords easy access to the top of the mountain range, a natural advantage exploited first by the railroad and more recently by the interstate highway.

Geologists include several formations in the Ogallala group, but because of the cut-and-fill stratigraphy, their spatial and temporal relationships are complex. Not all of the formations occur in any one place. Some are temporally equivalent to all or part of others in different locations. The Runningwater and Box Butte formations are exposed along the upper Niobrara River valley. A bit farther west and south in the Panhandle, a different set of formations are interwoven in the divide between the Niobrara and the North Platte Rivers—the Sheep Creek, Olcott, and Snake Creek formations. In north-central Nebraska, the lower Niobrara River exposes the Valentine formation, which is divided into three distinct members. The Ogallala sediments above the Valentine formation are the Ash Hollow formation. There are also many areas where the Ogallala group is not divided and is simply called the Ogallala formation, as it is traced down into Kansas.

Mammals and Other Life

Nebraska has produced a remarkable fossil record of the history of mammals, but it is just a part of their evolutionary story. Although the Cenozoic era, the latter half of which is well represented by sediments in Nebraska, is commonly referred to as the Age of Mammals, it does not encompass the history of that group. The evolutionary lineage that gave rise to the mammals has been around almost since the first vertebrates to live permanently on land, the amniotes. The earliest fossils of these four-legged animals, which laid eggs that could survive and develop on dry land, come from rocks of Pennsylvanian age. Sometimes called mammal-like reptiles because of the lack of specializations we associate with true mammals, this group of amniotes nevertheless had diverged down a distinct evolutionary path that leads only to mammals.

True mammals, which share many more features with the mammals we are familiar with today, first appear at about the same time as the earliest dinosaurs in Triassic time. The mammals lived throughout the reign of the dinosaurs in the Jurassic and Cretaceous periods, but they remained small, never much larger than a cat and usually much smaller. The dinosaurs' domination of the ecosystems of the world excluded the mammals from many ecological roles, especially those that took advantage of large size. But, marginalized as they were, the mammals evolved and diversified, so that most of the modern groups of mammals have their roots in Cretaceous time.

With the extinction of the dinosaurs at the end of Cretaceous time, the mammals were free to evolve to exploit the environment in ways that had not been open to them before. This exuberant diversification resulted in all manner of mammals, herbivorous and carnivorous, large and small, and they became the dominant land animals characteristic of the Age of Mammals.

In this outburst of adaptation, many of the orders that flourished did not have long-term success and became extinct. In addition, even among those orders that still thrive today, evolution wrought some drastic changes throughout the Cenozoic era. The combined effect of all this is that the mammals of the early Tertiary period, through Paleocene and early Eocene time, seem very peculiar to us. Large herbivorous beasts, such as pantodonts and uintatheres are among the archaic orders of mammals that disappeared with no descendants. Likewise, some of the large carnivorous mammals of that time, oxyaenids and hyaenodontids, persisted well into Tertiary time but have no living descendants. Meanwhile, the even-toed ungulates (artiodactyls) and odd-toed ungulates (perissodactyls), which account for most large, hoofed mammalian herbivores today, were just beginning their evolutionary success story and consisted of a few small species.

The early history of mammals has been pieced together from many areas, some as nearby as Wyoming and South Dakota, but is largely unknown in Nebraska. With the exception of a few early Eocene mammal teeth, so far we have no record of this early history. But the rich record of mammal fossils from the middle and later parts of Cenozoic time put Nebraska on the paleontological map.

Late Eocene and Oligocene time, which are represented in Nebraska by the White River and Arikaree groups, was a time of transition from the archaic mammal faunas of early Tertiary time to the modern mammals of late Tertiary time. Some of the doomed orders were still quite successful but were becoming outnumbered by representatives of groups that are more familiar to us. For example, *Hyaenodon,* one of the last representatives of an early group of carnivorous mammals, is one of the largest predators found in the White River group. But living alongside them were early representatives of modern carnivores—*Hesperocyon,* the earliest canid (dog) and *Pseudaelurus,* an early felid (cat). Even within the modern orders, though, some evolutionary branches did not continue to the present. The hoofed, herbivorous mammals followed a similar pattern. *Mesohippus,* a three-toed horse, and primitive rhinos and tapirs represented the odd-toed ungulate groups we know today, but brontotheres and chalicotheres went extinct.

Throughout Miocene time, extinction and evolution transformed the mammalian faunas of the world, eliminating the archaic groups and modifying the survivors towards their present forms. Fossils of the Ogallala group reveal much of this transformation. If you were transported back to the end of Miocene time, you would probably be able to recognize most of the mammals to some extent, at least to which modern group each belonged. Mammal faunas of the Pliocene and Pleistocene were very like those of the modern world but with greater diversity. Some of the biggest changes from Pleistocene time to the present have been extinctions, especially of the large mammal species, the megafauna.

White River Fossils: The fauna of the Chadron formation of Eocene time includes some animals that are absent from the Brule formation; it appears they didn't survive into the dryer climate of Oligocene time. This includes alligators and huge browsing mammals known as brontotheres, which were the last of their kind in North America. Though much larger, they probably resembled modern rhinoceroses; from the fossil record we know that they were odd-toed ungulates. Brontotheres had a large saddle-shaped head that bore a pair of blunt, bony protuberances in front. Males may have used these hornlike structures to compete for mates.

Odd-toed ungulates were the dominant large herbivores during Eocene and Oligocene time, and the White River group includes fossils of a number of different families. Medium-to-large, hornless, archaic rhinoceroses such as *Hyracodon* were common. Tapirs, close kin to the rhinos and their modern namesake, lived in Nebraska, along with the early horse, *Mesohippus.* It was the size of a large dog and, unlike modern horses, had three functional toes on each of its feet.

The even-toed ungulates in Eocene and Oligocene time were primarily small herbivores. Among the most common mammals in the fauna were oreodonts, often described as piglike or sheeplike, though neither description provides an accurate picture of the animal. It had teeth more like those of sheep than like those of pigs, but its skull was broad and its skeleton, with short legs and

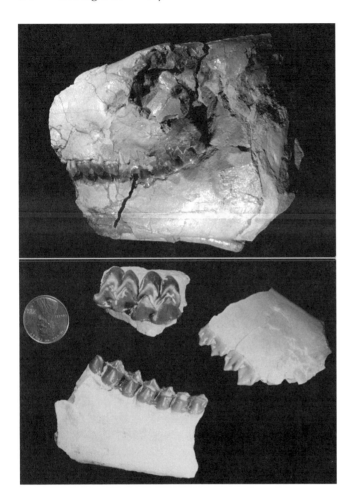

Crushed oreodont skull with lower jaw about 6 inches long *(top)* and oreodont jaw fragments with teeth *(bottom),* from the Brule formation. The skull is damaged so that the front of the snout *(left)* and much of the braincase *(right)* are missing. The position of the eye is indicated by the crushed orbit in the top center of the skull.

a long tail, was not highly specialized for running as were many of its contemporaries. Camels, comparable in size to the contemporary horse, and small ruminants, relatives of deer and antelope, were other even-toed ungulates that lived at this time. Although less common, there were large even-toed ungulates such as the long-faced anthracotheres and the piglike entelodonts. And, speaking of pigs, the early peccary *Perchoerus* represents the beginnings of that uniquely North American group.

Very small mammals were also abundant; the most common fossils belonged to rabbits. Rodents ranging from mouse size to squirrel size, a small opossum-like marsupial, and a peculiar, hopping, insectivorous mammal lived in the woodlands as well. Lizards lived alongside the small mammals.

Tortoise shell embedded in Brule formation volcanic silts. Camera lens cap rests on sediment that fills the shell cavity. Note the flat base and concave surface of the shell cavity. Shell pieces like the one at the far left are the most common fossil occurrence in the Brule formation. Note the grooves on the shell surface where scales, or scutes, were attached to the bone. Shrink-swell clays, derived from the alteration of volcanic ash, created the surface texture of cracks, informally known as popcorn, in the surrounding sediment.

The carnivorous mammals of the Oligocene include archaic lineages that were headed for extinction, as well as early members of modern groups. The largest predator of the White River fauna was the largest of the species of *Hyaenodon,* a wolf-size animal with a long powerful muzzle full of teeth made for slicing and puncturing. Only slightly smaller were the amphicyonids, which, despite their doglike appearance, were early members of the lineage that produced bears. The earliest true dogs are also present in the White River fauna, but they were smaller carnivores, only about the size of a fox. The branch of carnivore evolution that leads to cats is also well represented by the nimravids, archaic relatives of true cats, and the earliest ancestors of the true cats themselves. Some of the nimravids developed the greatly enlarged upper canine teeth that we refer to as saber-toothed. To find such diversity of predators and their prey species today, you would have to look to the African veldt.

Fossil horse from the Valentine formation in the Ogallala group on exhibit at the University of Nebraska State Museum in Lincoln.

Although the abundant and distinctive mammalian fauna made the White River group world famous among paleontologists, the most common vertebrate fossils are not mammals at all, they are tortoises. John Clark, a paleontologist who did field work in the White River group before collectors removed most of the fossils from the surface, once said there were White River group exposures where one could walk across a large area by stepping from one tortoise shell to another.

Arikaree Fossils: Fossils of mammals have been found throughout the Arikaree group but are especially abundant from quarries at Agate Fossil Beds National Monument. Many different kinds of herbivores lived on the Nebraska savanna, browsing in wooded areas or grazing on the grasslands, a relatively new feature on the earth because grasses had just evolved. One of the most common fossils in the quarries is the small rhinoceros *Menoceras,* which had a pair of small horns side-by-side at the end of its nose. Horses were also common on the savanna and included both browsing and grazing species, a distinction paleontologists have interpreted from the structure of fossil teeth. Both had three-toed feet and were a bit larger than the horses from the White River beds, but the grazing horses had larger, taller teeth that could better withstand the heavy wear from a diet of grass.

Along with these familiar types of odd-toed ungulates, there were some peculiar relatives unlike any that survive today. The chalicothere *Moropus,* a browsing odd-toed ungulate, was the size of a modern horse but had front legs that were much longer than its hind legs and it had large claws rather than hooves. How it used those claws is a question that has puzzled many paleontologists.

The even-toed ungulates are well represented in the Arikaree fauna. Oreodonts related to those in the White River beds were moderately diverse and common, filling a variety of ecological niches. Another holdover from the earlier fauna is a giant, piglike entelodont, similar to a modern buffalo in size and proportions but with a long knobby head that carried canine teeth the size of bananas. The camel group had diversified and included a tall, long-necked species as well as a small, graceful species whose form resembled that of a gazelle. An extinct relative of the camel, *Syndyoceras* was a ruminant with bony, hornlike structures growing out of its skull.

All of the predators of the Arikaree fauna were true carnivores, belonging to the modern order Carnivora. The doglike amphicyonids were among the largest carnivores and are informally known as bear-dogs. They denned in burrows along the river valleys. Fossils of true dogs (canids) have also been found in the Arikaree group. As in the White River fauna, true cats (felids) existed alongside their close relatives, the nimravids, some of which were saber-toothed cats. Small carnivores of the weasel family also hunted for prey in western Nebraska in mid-Tertiary time.

Ogallala Fossils: During late Tertiary time when the Ogallala group was deposited, Nebraska was a vast grassland with abundant, diverse wildlife, perhaps not very different from the veldt ecosystem that is rapidly vanishing from modern-day Africa. Hackberry seeds, yucca roots, and grass seeds are commonly preserved in the ancient soils of the Ogallala, suggesting a climate similar to today's, although perhaps a bit drier seasonally. In some places, opal has replaced and fossilized plant structures in exquisite detail, even preserving the cell structure.

A diverse array of ruminants grazed on the grasses of the plains or browsed on vegetation along the rivers. Several different kinds of horses dominated the fauna, and those that specialized in open-country grazing evolved to have just one functional toe on each foot. Along with the horses, there were rhinoceroses. Some large ones resembled modern rhinos, but one very common species, *Teleoceras,* was short-legged and barrel-chested, resembling a hippopotamus. These rhinos are well represented at Ashfall Fossil Beds State Historical Park.

Intercontinental Exchange via Land Bridges

Not all of the abundant animal life that inhabited Nebraska during Tertiary time was homegrown. Immigrants from Eurasia repeatedly entered the ecosystem. For example, the fauna of the White River group includes fossils of felids, the true cats that have no previous history in North America. And, among the fossils of the Arikaree sediments, the big chalicothere *Moropus* and some of the carnivores such as the amphicyonids show connections to the Old World. Proboscideans, the group that includes modern elephants, are an ancient order of mammals with a long history in Africa and Asia but are absent from the fossil record in North America until early Miocene time, when they show up in Ogallala sediments. The mammoth, another proboscidean, probably originated in the Old World before immigrating to North America. Its appearance in the fossil record is used as an indicator of the beginning of Quaternary time.

The pathway of immigration was not a one-way street. At the same time that immigrants from Eurasia were arriving in North America, species that evolved here were finding their way into the Old World. The equids, horses in the broad sense, evolved in North America, but some of the diverse lineages of horses spread to Asia at various times during Tertiary time. Even *Equus,* the genus that includes all modern species of horses, appears to have evolved in North America and then spread throughout the world.

Skull of a camel from Pliocene sediments in Garden County, Nebraska, on exhibit at the University of Nebraska State Museum in Lincoln.

The extinction of large mammals in North America at the end of Pleistocene time claimed all the horses. It is ironic that historians often say that the Spanish, upon arriving in the Western Hemisphere and finding no native horses, introduced the horse to North America. They brought horses bred from the lineages of those original North American horses that had survived in Eurasia. Camel fossils from North America tell a similar story; ancestors of the modern camel evolved for most of their history in North America and spread to the Old World and South America late in Miocene time. Then they went extinct in North America, surviving only in the places they had emigrated to.

What is the explanation for all this traffic? North America must have been connected by land to Eurasia. It seems likely that this connection was at the Bering Strait where, today, only a narrow shallow sea separates Asia from North America. The Bering Sea floods a continental shelf that has connected North America and Asia ever since tectonic plate motions brought them together sometime in Cretaceous time. When sea level is as high as it is today, the low margins of the continents are flooded and a sea gap of varying width separates the two continents. But when sea level is low, as it is when much water is frozen within the polar ice caps, a strip of land emerges, growing broader as sea level falls. Just as the varying size of the south polar ice cap may have controlled fluctuating sea levels in Paleozoic time, it may have done so throughout Tertiary time. Whenever the land emerged above sea level, some animals migrated from one continent to the other.

Some groups never took advantage of this opportunity. The antilocaprids, the group that includes the pronghorn antelope, either didn't cross the strip of land or didn't survive long after it did. The pronghorn is not a true antelope like those that inhabit Africa but is actually the last of a group of ruminants that has been restricted to North America throughout its entire evolutionary history. Why didn't groups like the antilocaprids cross over? Perhaps suitable habitat did not exist all along the way or perhaps similar species already occupied suitable habitat on the other continent and prevented the newcomers from establishing themselves.

Plate tectonics and sea level changes have formed and broken many such land bridges throughout the history of the earth, but as far as Nebraska is concerned, the Bering Strait route had the greatest impact. There also may have been a direct connection to Europe across the North Atlantic until early Tertiary time. It became progressively more tenuous—finally breaking—as the North Atlantic opened. A South American connection to North America seems to have formed late in Miocene time, but much of the traffic traveled south. One immigrant group from South America that shows up in the fossil record of Nebraska in late Tertiary and Quaternary time is the edentates—giant ground sloths and armadillos.

Pliocene-Pleistocene Gravels

A series of coarse gravels deposited in river channels of Pliocene and early Pleistocene age in western and central Nebraska mark a major transition from net deposition to net erosion on the east side of the Rockies. The dominance of erosion and channel cutting by rivers and the preservation of isolated, coarse channel deposits has continued throughout Quaternary time. These gravels lie lower than, and cut into, the Gangplank surface but are significantly higher than the Platte River drainage and young Pleistocene gravels. They represent the initial dissection into the Gangplank surface.

Unlike the myriad channels in the underlying Tertiary Ogallala group, with finer-grained sediments from closer sources, these gravels consist of metamorphic and igneous clasts clearly derived from the Rocky Mountains to the west. Some of these gravels contain significant amounts of anorthosite, a rare igneous rock composed mainly of plagioclase feldspar. The source for these gravels is the Laramie Mountains in Wyoming, where a large body of Precambrian anorthosite is exposed.

These gravels include the Broadwater formation, examples of which perch on benches on the north side of the North Platte River valley and elsewhere in western Nebraska. An unconformity, a gap of perhaps a million years or more, separates the underlying Ash Hollow formation of the Ogallala group from the Broadwater gravels. In north-central Nebraska, gravels of the Long Pine formation, Pliocene and Pleistocene in age, are well exposed near the town of Long Pine.

These Pliocene-Pleistocene gravels occur on the tops of ridges, which indicates they are not associated with the present river system. A good example is the Pumpkin Creek drainage. The slopes above Pumpkin Creek, which drains Wildcat Ridge, contain these gravels, yet the present Pumpkin Creek drainage doesn't reach anywhere near the Laramie Mountains. Other examples of the anorthosite-bearing gravels are near the modern Republican River in south-central Nebraska. These deposits indicate that a vigorous Pliocene-Pleistocene river system, with its headwaters at least partly in the Laramie Mountains, used to cross Nebraska.

Because of a process known as topographic inversion, we see some of the old Pliocene-Pleistocene channel deposits high up on ridges. The channel deposits represent the lowest points in the valley of their time, but these channel gravels are more resistant to erosion and they armored the finer material below them from erosional forces. As the surrounding, more easily eroded deposits were removed, the channels remained and are preserved high in the landscape.

Quaternary Geology and the Modern Landscape

The Quaternary period runs from about 2 million years ago to the present. Most of Quaternary time falls within the Pleistocene epoch.

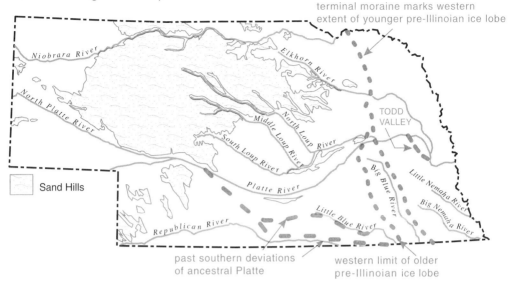

Geologic features from Quaternary time: modern river drainages, the limit of glacial deposits, the distribution of dune deposits, and the position of some old river channels.

Only the last 10,000 years, during which conditions on the earth have been much as we find them today, fall within the Holocene epoch.

Quaternary time in Nebraska encompasses the accumulation of extensive surficial deposits by a variety of sedimentary processes, the development of diverse landforms, a record of great climatic variation, the migration and settlement of diverse peoples, and present-day environmental concerns. Most of the state is covered by Quaternary deposits. For those geologists who focus on the Quaternary period, Nebraska is a gold mine of unique features and an archive of interesting events: Quaternary sediments record the work of ice sheets, dust storms, migrating dunes, and errant rivers. Old floodplains, preserved above the active floodplains, highlight the local cycles of downcutting. Debris piles left by Ice Age glaciers mantle the eastern part of the state, and a sand sea lies dormant over one-third of the state.

Ice and the Missouri River Valley

Throughout eastern Nebraska one occasionally finds isolated boulders, some of which are quite large and end up in front yard landscape projects. These boulders are called erratics because they are not from the local bedrock—they wandered considerably in their journey to Nebraska. Many of the erratics are a very hard pink quartzite, a distinctive metamorphosed quartz sandstone. The closest surface exposures of this rock, known as the Sioux quartzite of Precambrian age, occur near the point where Iowa, South Dakota, and Minnesota meet. Erratics of Precambrian granites and greenstones are also common in Nebraska; they were transported from the Canadian Shield, an even more distant source of ancient basement rocks.

What transported these boulders hundreds of miles from their bedrock sources? Erratics are part of a distinctive glacial deposit, known as till, that consists of particles that range in size from clay to boulders. Other evidence of glaciers in Nebraska include striations—parallel scratches made by harder particles (large and small) dragged by the glacial ice over the underlying bedrock; animal and plant fossils characteristic of cold climates; and some moraines.

Glaciers are large, moving bodies of ice. They form over many years wherever the amount of snow that accumulates during each year is greater than the amount that melts. If the accumulated ice is thick enough, the densely compacted ice at the bottom of the glacier begins to flow. Glaciers form on high mountains and flow down valleys, growing or receding depending on the climate. But where vast sheets of ice, sometimes miles thick, form over a large area, the ice flows outward from the center of accumulation, regardless of the slope of the land over which it moves. The largest such continental glacier existing today covers the continent of Antarctica. During Pleistocene time, vast ice sheets also formed in the Northern Hemisphere and spread southward. These extensive glaciers, some of which reached all the way from Canada into Nebraska, formed during periods of profound global climate change—the Pleistocene ice ages.

As ice flows over the landscape, it is an extremely powerful erosional force, scouring loose material, plucking fragments from more resistant rock. But along the margins of the glacier, where the ice melts faster than it flows out, the sediment carried by the ice is simply dropped.

Till, the debris carried by the ice, is dropped at the margin of the glacier as the ice melts. Accumulations of till are called moraines. Terminal moraines delineate the farthest advance of the ice sheets, and ground moraines cover the ground over which the glaciers retreated.

Geologists used to think the Pleistocene ice ages consisted of four stages, from youngest to oldest the Wisconsinan, Illinoian, Kansan, and Nebraskan. Tills of at least two different ice ages—the Kansan and Nebraskan—are present in Nebraska, but subsequent work has shown that these two simple divisions do not adequately represent the early glacial advances in the continental interior. Some geologists have suggested that the terms *Kansan* and *Nebraskan* be abandoned, and that we should refer to these more complicated early glacial advances as *pre-Illinoian*. Future research will help to unravel this history. Above the oldest glacial deposits in Nebraska are fluvial outwash sediments, which in turn have volcanic ash deposits in them that are up to 1.2 million years old.

Large fingers, called ice lobes, protrude farther south from the ice sheet where ice flow was faster. A pre-Illinoian ice lobe extended down the axis of the Missouri River all the way into Kansas, leaving glacial deposits in the eastern part of Nebraska. A moraine that runs north-south through eastern Nebraska marks the western margin of the ice lobe. Most of the moraine has been destroyed or highly degraded by various geologic processes. The rest of Nebraska—west of the moraine—was never glaciated. The

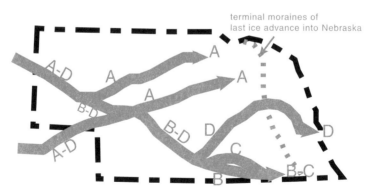

A - Late Pliocene (about 2,500,000 years ago)
B - Early Pleistocene (about 1,500,000 years ago)
C - Illinoian (about 200,000 years ago)
D - Late Wisconsinan (about 30,000 years ago)

Model for how Platte River drainage changed with time. Early Ice Age glacial advances deflected the Platte River to the south. —*Modified from Souders, Swinehart, and Dreeszen, 1990*

modern Missouri River likely inherited its position from the retreating ice lobe. Buried valleys, identified from drill hole data,
suggest the river drainage prior to and during the beginning of
Pleistocene time was notably different, flowing westerly in eastern
Nebraska and western Iowa across the present Missouri River
drainage system. Pre-Illinoian ice lobes redirected this drainage to
the south. When they retreated, the Missouri River followed the
glacial topography.

The upper part of the Missouri River was also shaped by the
Wisconsinan glaciation. An ice lobe extended south to the northeast Nebraska–southeast South Dakota border. The Missouri River
followed the western margin of the lobe down through South Dakota
and then curved east, following the southern edge of this lobe at
the Nebraska border.

Between the major glacial advances of Pleistocene time, there
were warmer interglacial intervals. The duration and effects of
climatic oscillations vary quite a bit. The earth has experienced
many mini–ice ages, even within the past 10,000 years, and some
climatologists argue that we are in an interglacial interval now; in
time, the glaciers may return.

Where Did All the Mammoths Go?

A megafauna of large mammals, including mammoths, lions, saber-
toothed cats, musk oxen, and camels used to walk the Nebraska
plains in Quaternary time. One species, *Bison latifrons,* had a 7-foot
horn span and was 25 percent larger than the modern bison. A
classic signature fossil of the time is the mammoth. Mammoth
remains have been found at well over one thousand sites in Nebraska and from almost every county. In a spectacular find near
Crawford, in northwest Nebraska, two male mammoth skeletons
were found with the tusks interlocked. Standing some 13 feet tall
at the shoulder, they died sparring for territory or mates. Underneath one was a crushed coyote skull, suggesting a grisly end even
for one scavenger.

Mammoths and many other mammals of the megafauna went
extinct at the end of the Pleistocene Ice Age. Scientists continue to
debate the cause of these extinctions. Clearly, the mammoths
and other Ice Age fauna were adept at living near the ice sheets;
their demise may have been related to the warming climate and

Mural of mammoths from elephant hall at the University of Nebraska State Museum in Lincoln. *—Art by Mark E. Marcuson. Reproduced with permission from the University of Nebraska State Museum, the copyright holder*

changing vegetation. However, they survived many earlier climatic fluctuations. The appearance of humans at the time of these extinctions led scientists to speculate that the mammoths and other megafauna were overhunted. The Hudson-Meng bone bed, which entombs the densely packed bones of six hundred bison, may be the result of this type of hunting. Clear evidence indicates humans were living in Nebraska 9,500 years ago as indicated by a style of projectiles known as Alberta points found at the Hudson-Meng site. Possibly older projectile points of the prehistoric Clovis culture have been found loose on the surface, but their age cannot be definitely determined. Bison and bighorn sheep are among the few surviving megafauna from this time. The bison evolved to their present smaller size.

Dust of the Plains

Along the Missouri River in the Omaha region are distinctive bluffs composed of massive light-tan silt deposits, known as loess, that are up to several hundred feet thick. The loess, though easily crumbled by hand, is weakly cemented and has the strength to maintain the steep slopes of the bluffs. While remarkable for its relative homogeneity, the loess is composed of several different units with somewhat diffuse boundaries and differences in colorations and clay content. At certain levels, snail fossils are common in the loess, as are calcareous concretions known locally as loess *kindchen,* German for "children," because they resemble the heads of young children. Formed from groundwater action, the concretions, ranging from pea to grapefruit size, are commonly cracked and hollow inside and sometimes have small crystals lining their interior. Elsewhere, old calcified root paths are preserved. Subvertical fractures characterize many exposures.

These distinctive silty deposits are windblown suspension deposits—simply accumulations of dust that settled out of the air. The ability of wind to carry large volumes of silt may not be appreciated by North Americans, but a fine layer of silt regularly covers everything in Peking, China, during the dry winters and after windstorms that blow from the deserts to the west. Multiply that fine layer by tens of thousands of years or more, and the accumulation of hundreds of feet of dust seems plausible. Geologists have found silt grains from the Sahara Desert—where the strong winds sometimes blow from the east—at the bottom of the middle of the Atlantic Ocean and even as far away as the Bahamas, which attests to the distances that windblown material can travel.

What trapped and allowed the dust to accumulate in Nebraska's loess deposits? You might think the next wind would just blow it away. One explanation is the baffle effect of grasslands—whereby silt that filtered down through existing grass cover was protected from subsequent winds. This dust-trapping mechanism is obvious to anyone who has cleaned an old carpet. Miners use a similar baffle mechanism to sort fine dense gold from other sediment; they wash the mix over a carpet, which traps the heavier gold while the other lighter sediments wash away.

With the semicontinuous input of loess, the grasslands grew upward. Soil-forming processes helped cement the loess grains together. When dust input was lower, soils developed to a greater degree, forming diffuse layers and boundaries within the loess. Wind does not necessarily deposit loess in a flat layer like water deposits sediment. Windblown deposits mantle the existing topography; in numerous places loess mantles older hills.

Loess deposits, well known in China, Hungary, and the interior of the United States, are associated with grasslands that border deserts that serve as source areas for the dust. Iowa is widely known as the location of the scenic Loess Hills along the eastern side of the Missouri River, but if Nebraska were to designate a state sedimentary deposit, a good case could be made for loess. In addition to deposits along the Missouri River from Tekamah south to Kansas, there are also extensive deposits in south-central Nebraska.

The loess supports a distinctive landscape. Steep-sided loess canyons and gullies are preferentially aligned in a northwest-

Younger loess mantles an older loess hill with a truncated, dark soil horizon developed in it *(just above vehicle)*. Photo from Broken Bow area.

southeast direction. Underground erosion, known as piping, plays a role in the formation of these canyons; groundwater moving along fractures or root systems removes the silt at the heads of gullies. Then, collapse and surface erosion increase the size of gullies and canyons.

The loess deposits in Nebraska are of several different ages. Three major units are the younger Peoria, the Gilman Canyon, and the older Loveland loess. Geologists have dated the organic material at the base and top of the Peoria loess at 23,000 and 13,000 years before present, respectively. In places it is more than 120 feet thick. This loess is more calcareous and lighter colored than underlying material. The Gilman Canyon loess, a thinner darker unit, was deposited 40,000 to 24,000 years ago. Because smaller amounts of loess accumulated during a longer time interval, we can infer a lower rate of loess accumulation at this time. Old soil horizons are also better developed in the Gilman Canyon loess.

The Loveland loess was deposited about 140,000 to 135,000 years ago and is one of the most widespread and thickest loess units in Nebraska. Typically yellowish to reddish brown, the reddish pigmentation increases towards the top of the unit. In places, a well-developed buried soil, named for the Sangamon interglacial stage that preceded the final Pleistocene ice advance, separates the Gilman Canyon and Loveland loess. It attests to a long period

Loess outcrops and canyon just southwest of Broken Bow.

of stability before loess conditions returned. Periods of loess generation separated by those of stability reflect regional climate changes.

Loess demands unique engineering solutions. Its relatively low crushing strength requires deep supports for larger buildings; it loses strength and its erodability increases upon reworking with tools such as bulldozers. However, farmers enjoy rich soils that formed on the loess-covered tablelands.

The Sand Hills

Covering almost 20,000 square miles in west-central Nebraska is a truly unique landscape known as the Sand Hills. Here, hills of myriad distinctive, smooth shapes up to hundreds of feet high are composed of sand held in place by a thin cover of grass. You can easily see the remarkably homogenous sand through the thin soil. In some places, distinctive bowl-shaped depressions tens of feet across, known as blowouts, expose bare sand.

Large parts of this landscape lack a drainage network—the branching pattern of small gullies feeding larger streams. Only a few long, linear rivers cut through the hills. In between some of the hills are flat depressions dotted by lakes, marshes, and meadows. The roots of the rich vegetation tap directly into a shallow groundwater table. The wetlands support a rich ecosystem—especially birds—including pelicans. Where buffalo and antelope used to range, the Sand Hills now support large cattle ranches. Few humans live in the region—about 17,000 people in 20,000 square miles.

These hills represent the largest sand dune field in the Americas. The growth of vegetation associated with wetter times stabilized the dunes. Today's annual rainfall of 18 inches maintains the vegetation. Such large sand dune fields are known as ergs, or sand seas. Driving through the undulating landscape the metaphor of a sea seems apt.

Dunes come in a great variety of forms, and seven varieties are found in the Sand Hills: barchan dunes, barchanoid-ridge dunes, crescent-shaped parabolic dunes, domes, domal-ridge dunes, linear dunes (elongate parallel to the prevailing wind direction), and sand sheets. Each form of sand body in the Sand Hills occurs in distinct patches within a large-scale pattern; this suggests their distribution is not haphazard. However, the factors involved are

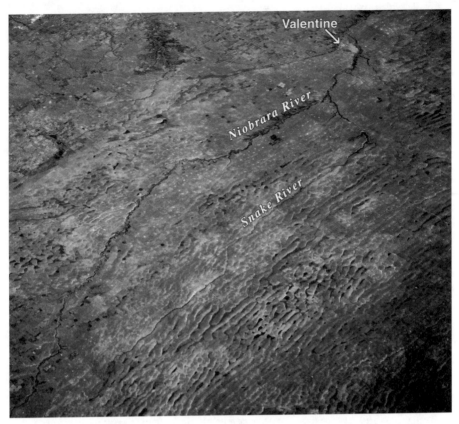

Oblique satellite image of north-central Nebraska looking northeast into South Dakota. Valentine is in the upper right corner. The long linear sand dunes with darker interdune depressions appear as stripes in the lower half. Dune geometry varies in different areas. You can trace the Niobrara River from where it bends near Valentine to the bottom left corner. To the south of the Niobrara, you can see the Snake River, which follows the dune orientation.

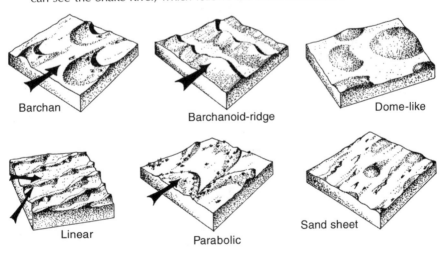

Types of sand dunes. —*From McKee, 1979*

complex and the answers not simple. Wind direction, consistency, and speed, the anchoring role of vegetation, and the amount of sand available all determine the sand dune form. The dune morphologies and internal structures indicate the prevailing winds were out of the north or northwest when the dunes formed.

Where the distinctly crescent-shaped barchan dunes are so close together that they coalesce, they build into a barchanoid-ridge dune perpendicular to the prevailing wind direction. Barchanoid-ridge dunes have a steep face—almost a 30-degree inclination—on the downwind side, and a shallow, more irregular face on the upwind side. They are some of the largest dunes in the Sand Hills, more than 250 feet higher than the adjacent, flat interdune depression areas, with the largest one more than 400 feet high. They range from miles to tens of miles long. They are especially common and striking in the northwestern part of the Sand Hills.

Large, well-formed barchanoid-ridge dunes look like giant ripples on satellite images. This has led two people, Hansen and Kelly, to suggest that the Sand Hills are not dune deposits but are actually the result of a giant wave, or tsunami, triggered by a large meteorite impact up north somewhere. This is geofantasy of an absurd type. Although it may seem to be consistent with a casual look at a satellite image, it is inconsistent with many other scientific observations. For instance, the hypothesis does not explain how a catastrophic wave could deposit such well-sorted sand. Such junk science may beguile the unwary and gullible, but evidence available to anyone can easily refute it.

The flat areas between the dunes formed in several ways. The naturally broad troughs between dunes may expose the upper surface of the Ogallala group or other sediments over which the sand dunes moved. Or they can represent places where the wind has eroded the sand down to the water table and no farther. The wetted grains at and below the water table are held together by capillary forces and are not easily blown away. Finally, the flat depressions could be areas of deposition where lakes and wetlands trapped the windblown silt and sand. In this case, the sediment would build up to and be protected by the groundwater level that determined the original lake's level.

Since the sand dunes rest upon loess deposits in places, they are clearly younger than the loess. The fact that their forms are

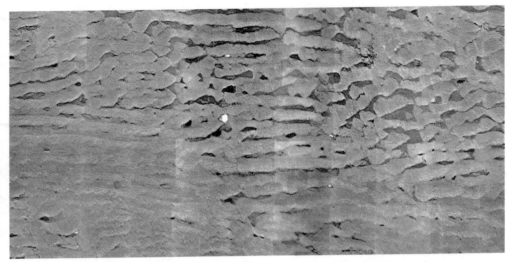

Composite air photo of Sand Hills some 15 miles north of Whitman shows large east-west oriented barchanoid-ridge dunes. The dark areas are interdune depressions—wetlands and lakes. Image is about 30 miles wide. —*Photo courtesy of U.S. Geological Survey*

well preserved also speaks to their youth. Jim Swinehart, David Loope, and colleagues at the University of Nebraska at Lincoln have expended considerable effort trying to determine their age. Coring through the dunes, the scientists retrieved organic material mixed in with the underlying sediments. The radiocarbon ages from this organic material gave the scientists an idea of when the dunes must have moved over the material. Some eighteen radiocarbon ages reveal that the dunes moved several times. The oldest material covered by dunes is about 15,000 years old and the youngest is a mere 800 to 900 years old. Taken in total they suggest that dune formation started some 14,000 to 11,000 years ago, at the end of the Wisconsinan—the last ice advance. There was another major period of dune activity from 8,000 to 5,000 years ago. After roughly 1,000 years of stability, the sand dunes moved again, followed by another period of stabilization about 1,500 years ago. From a geologic perspective, the Sand Hills are merely dormant and have the potential to move again.

It is possible that an extended drought or climatic shift could remobilize the dunes by destroying the vegetation that holds them in place. Active blowouts remind us of this potential. In the present climate, colonizing plants, such as blowout grass, quickly establish a foothold and heal these depressions in a matter of decades. But

during the drought of the 1930s, active blowouts were more numerous. The magnitude of drought or other environmental change
necessary to cause the dunes to become active again is not known,
but clearly the dunes have remobilized at times, including some
activity in the past 1,000 years.

Human activities also induce blowouts. Small blowouts form
around telephone poles or fence posts, and overgrazing by cattle
can create them. Some ranchers cover blowouts with tires in an
attempt to slow erosion. Take notice of the vegetation alongside
the roads in the Sand Hills. It tends to be denser and includes
more species—often wildflowers—than the grass cover beyond the
fences where grazing occurs. Clearly a complex interplay between
geologic and biologic forces—including human activity—and climatic history has shaped and will continue to shape this delicate
landscape.

Stream drainage networks do not form within the Sand Hills
because rainfall immediately seeps into the very porous and permeable dune sand—an effective sponge. A high percentage of the
rainfall thus goes directly into replenishing the groundwater. The
Sand Hills, a large pile of water-saturated sand, is a major, high-
quality aquifer, which feeds the High Plains aquifer in the Ogallala
sediments below. Seepage out of the sands feeds the many lakes,
marshes, and meadows. This shallow groundwater table supports
a diverse surface life. The local ranchers use the meadows for
haying and grazing, and wildlife, especially waterfowl, use the
wetland habitat. Groundwater seepage also feeds the branches of
the Loup and Dismal Rivers, providing a relatively constant water
flow despite the lack of a supporting drainage network. Ninety
percent of the flow in the Dismal River comes from groundwater.
Few other landscapes in the United States depend so much on the
groundwater. Extensive groundwater withdrawal could locally lower
the water table enough to destroy meadows and wetlands.

When the dunes were on the march they affected rivers. This
became evident after geologist Robert Diffendal found lake sediments perched on the slopes above present-day Lake McConaughy
on the Platte River. The sediments were radiocarbon dated to about
13,000 years ago, and given their elevated position they suggested
that a lake or cluster of lakes much larger than Lake McConaughy
occupied the North Platte River valley. This old lake was subse-

quently named Lake Diffendal to honor its discoverer. Sand dune migration likely dammed the river, creating Lake Diffendal, a story we describe in greater detail in the Ash Hollow geolocality. Since this initial discovery, geologists have discovered that dune migration has formed lakes in other drainage basins in the Sand Hills.

Groundwater

Vast amounts of groundwater, an important resource, move through the pore spaces and fractures in rocks and sediments. Geologic units that can store and transmit significant amounts of water are known as aquifers. A good sandstone aquifer can easily be one-third water by volume. Groundwater moves through these aquifers at a relatively slow pace.

Rocks that serve as barriers to underground water flow are known as aquicludes. The amount and interconnectedness of pore spaces and fractures in the rock determines whether it is an aquifer, aquiclude, or something in between. Glacial deposits can often be aquicludes, but gravels associated with the meltwater rivers of the same glaciers can be good aquifers.

Groundwater is connected to surface water by springs, seeps, marshes, lakes, and meadows, all part of the hydrologic cycle. Sometimes surface water seeps down to the aquifer recharging it and sometimes the groundwater discharges to the surface. Understanding this connection, the science of geohydrology, is critical for managing water resources. Some ecosystems in Nebraska are dependent on surface-groundwater connections. Nebraska, a very interesting place to study groundwater behavior, has some very large aquifers—much more water is below ground than above ground.

River deposits are often good groundwater aquifers, especially well-sorted sand and gravel. In the western Platte River floodplain, farmers irrigate their crops using groundwater pumped through center-pivot irrigation systems. A well is at the center of each pivot. Omaha also gets about half of its water from a well field in the Platte River alluvium, and the city plans to develop another well system farther upstream. The groundwater in this alluvium has a direct connection to the water flowing aboveground in the river. Unfortunately, groundwater in the Platte River corridor is also locally contaminated with nitrates, atrazine, and other agricultural compounds.

Rivers Running Amok

More than one person driving to the airport in Omaha has wondered why a small piece of Iowa, encircled by Carter Lake, is on the west side of the Missouri River in what should be Nebraska. This happened because rivers are more dynamic than state boundaries. The U-shaped Carter Lake used to be part of the Missouri River channel, and the land inside the channel loop was on the east side of the river. The Missouri River abandoned the channel after the initial state boundaries were drawn, so a small piece of Iowa now lies west of the river.

Understanding a bit about how rivers behave is not only important to understanding some political boundaries but crucial to understanding Nebraska's geology. Rivers are major landscape features and have been throughout Quaternary and much of Tertiary time. They are responsible for a significant percentage of Nebraska's surficial deposits. Major east-west drainage systems—the Niobrara, Elkhorn, Loup, Platte, Nemaha, and Republican Rivers—all feed the Missouri. Their character is quite diverse, including meandering rivers on broad flat floodplains, braided rivers choked with sediment, and rivers that have cut canyons.

The river systems changed dramatically during Quaternary time with the coming and going of the ice sheets. A former channel of the Platte is now a broad riverless depression known as the Todd Valley. An even earlier version of the Platte River flowed south into Kansas.

Where a river channel curves, the water flows faster on the outside of the curve and tends to erode the banks, while on the inside of the channel slower water tends to deposit some sediment, often as sandbars that build out from the bank. In this way the channel migrates in the direction of the bend. During floods, the river sometimes cuts across its floodplain and forms a new channel, abandoning the old channel, which often becomes a curved water body known as an oxbow lake. During normal flow conditions the currents mainly move sand and gravel, depositing it in the channel, but during floods, water spreads mud and silt in layers over the floodplain. The Elkhorn River is a good example of a largely unengineered river that still meanders; these movements leave curved patterns on the floodplain that are visible in air photos.

A 1993 air photo of the Elkhorn River near Scribner shows an abundance of meander bends with sandbar deposition on the interior of bends and oxbow lakes and meander scars on the floodplain. Width of photo area is about 4 miles. —*Photo courtesy of U.S. Geological Survey*

For any section of river channel, the system can be in one of three general states—it can be downcutting, stable, or building up. Which state the river is in is a complex function of many factors including the amount of water, the channel configuration (including the gradient), and the amount and type of sediment carried.

Fairly often, especially in western Nebraska, slightly higher flat areas edge the present river floodplain. Known as terraces, these old channel and floodplain deposits are the products of cut-and-fill cycles: the river cut down into its former floodplain, establishing a new floodplain at a lower level. Multiple terraces indicate a complex history of cut-and-fill cycles, which are often attributed to climate shifts, tectonic movements, and changes in sea level. Detailed studies of terraces help reconstruct geologic history. Nebraska went through a megaregional cut-and-fill cycle in Tertiary and Quaternary time when the rivers deposited and then eroded the Gangplank surface. Today, many small rivers in Nebraska are in a downcutting portion of the cycle and are entrenched, having cut down into floodplains they built quite recently. They now flow in steep, narrow channels well below these recent floodplains and have not yet had time to establish a new equilibrium with a new floodplain at the lower level.

The Missouri River used to wander all over its floodplain. Early steamboat pilots never knew exactly what to expect and often ran aground. In 1943 the sudden arrival of warm weather melted a 2-foot snowpack, increasing the river's discharge to about 200,000 cubic feet per second. It flooded the Omaha airport under 7 feet of water, and one thousand people had to evacuate their homes. In places the river was 15 miles wide. Total damage in 1943 dollars was estimated at 1.4 million. Major floods have occurred along this section of the Missouri River in 1881, 1943, 1952, 1967, 1978, and 1993. The 1952 flood was the largest with a discharge of 396,000 cubic feet per second. For comparison, the annual average flow is typically 30,000 to 50,000 cubic feet per second.

In the 1950s, the U.S. Army Corps of Engineers stabilized the river for flood protection and commercial purposes. They placed fingers of coarse rock debris that jut out from the shore and slant downstream at regular intervals along the channel. You can see these jetties at times of low water flow. They concentrate the flow in the channel center and prevent bank erosion. In addition, the Corps closed off secondary channels. Recently, however, the Corps reopened secondary channels to provide spawning and wildlife habitat, floodwater storage, and recreational opportunities. A good example of this is at Boyer Chute just north of Omaha.

Human activity has also significantly changed the Platte River. One hundred years ago it acted more like a typical braided river. It had a broad floodplain of bare sediments up to several kilometers wide with many shallow channels. Spring floods scoured the multitude of channels and kept vegetation from taking hold. The classic refrain was that the Platte was useless—too thick to drink and to thin to plow. We have constructed dams upstream to regulate, moderate, and decrease the flow so that spring floods no longer scour shifting sandbars, which have stabilized as wooded islands, and the number of channels and the extent of exposed sandbars has drastically diminished. In addition, lakes now dot many areas along the floodplain where groundwater has filled abandoned gravel quarries.

Scientists are concerned about the effect of such changes on wildlife habitat. Every spring, flocks of migrating sandhill cranes stop along the Platte River in central Nebraska. Here they conduct mating rituals and gain strength for the remaining journey north. Fossils of cranes in Nebraska suggest that this spring ritual has occurred for millions of years, first on the ancestral Platte and now on the modern river. The challenge for state and federal natural resource agencies is to apportion water among the competing demands of irrigation, power production, and wildlife habitat maintenance.

Nebraska has been engaged in water disputes with Colorado, Wyoming, and Kansas. Kansas took Nebraska to court for using too much of the Republican River's water before it crossed the border. Wyoming and Nebraska settled another suit in which Nebraska contended that Wyoming was taking too much water out of the North Platte. An unsettled question critical to resolving the legal disputes is how much wells adjacent to the river impact the stream flow. Although the courts have dealt with surface water and groundwater disputes separately, abundant scientific evidence indicates that the two sources are connected, and the use of one impacts the other.

Major devastating floods led to the construction of dams and levees along many rivers in Nebraska. Harlan County Dam was built after a disastrous flood in 1935 along the Republican River. Extensive levees were constructed along the Missouri River after flooding in 1943 in Omaha. Reservoirs provide flood control, recreational opportunities, and irrigation water, but they also change river dynamics and sedimentation patterns. Debate about water policy in Nebraska and in the western United States will only increase in the near future, and understanding the complex dynamics of rivers and groundwater behavior will become more important.

Interstate 80
Omaha—Wyoming
455 miles

Interstate 80, one of the most heavily traveled roads in the United States, has been called America's Main Street. Little wonder, since it connects New York, Chicago, and Denver. Many people drive through Nebraska on their way to someplace else and develop an impression of the state from what they see along the nearly 500 miles of this road. Unfortunately, this is one of the least interesting, if not outright boring, paths through Nebraska in terms of scenery and striking geological features. Like the transcontinental rail line it closely parallels, I-80 follows the route that presented the fewest obstacles to westward construction and travel.

Though you are probably traveling at 75 miles per hour, with a practiced eye and a little guidance, you can catch a glimpse of many subtle but important geologic features. Though vegetation covers most of the recent sediments, which in turn blanket the older rock layers along this route, I-80 follows the Platte, a dynamic shifting river, across Nebraska. You can also see evidence of several important environmental concerns along this route. We strongly encourage you to deviate from this beaten path. Several geolocalities are along or near I-80: Schramm Park on the Platte River between Omaha and Lincoln; the University of Nebraska State Museum in Lincoln; and Ash Hollow State Historical Park, about 25 miles west of Ogallala on U.S. 26.

On the eastern border of Nebraska, the I-80 bridge across the Missouri River carries the road well above the floodplain and delivers it to the top of the bluffs in South Omaha. These thick deposits of loess mantle the slightly older Pleistocene tills and much older Mesozoic and Paleozoic rocks. Small drainages dissect the bluffs, creating surprisingly steep topography. From the bridge, you can see Omaha's botanical gardens cut into the pale-colored loesses just to the north of I-80. For the first couple of miles in Nebraska, the interstate passes through the urban canyons of roadcuts and overpasses and emerges near exit 451 into a more open area to the south.

Most of the city of Omaha lies within the drainage basin of Big Papillion Creek, and I-80 crosses the valley of Little Papillion Creek

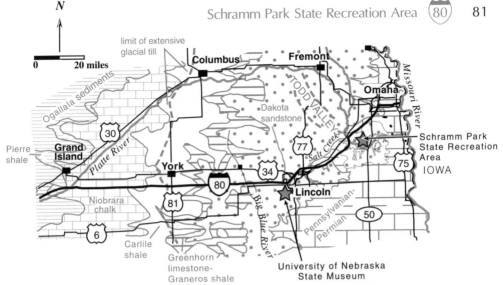

Geology along I-80 between Omaha and Grand Island. —*Modified from Burchett and others, 1972; Burchett and others, 1975; Burchett and Pabian, 1991*

between exits 450 and 449 and Big Papillion Creek at exit 448. These creeks drain to the south, turning east toward the Missouri River well south of the city. Omaha has dealt with the flood hazards posed by the creeks by designating floodways along Big Papillion Creek and its tributaries, in which development is limited to parks and trails. These drainages are channelized, with levees, riprap, and other drainage control measures. Another example of engineering solutions to flood control is Wehrspan Lake, just to the west of exit 440, one of several reservoirs created by flood control dams surrounding Omaha. In spite of these precautions, flash floods still damage parts of Omaha on rare occasions, most recently in 1999.

Schramm Park State Recreation Area

Schramm Park State Recreation Area, on the north bank of the lower reaches of the Platte River, is easily accessible from I-80; head south at exit 432. Most of the major geologic units in the Omaha area are exposed in this park. Starting at the south end of the park and extending for half a mile downstream along the edge of the Platte are intermittent cliffs of Kansas City group limestones and shales of Pennsylvanian age that are tens of feet high. You can find fossils of fusulinids, crinoids, and corals.

At the south entrance to the park, a display at the base of a cliff describes the geology. The shales exposed in the cliff are reddish and green. An overlying disconformity and weathering associated with Cretaceous times likely creates some of the coloration. Though it is not well exposed in the cliff, you can find large blocks of brown, iron-cemented sandstone of the Cretaceous Dakota formation at the foot of the slope. The disconformity between Pennsylvanian and Cretaceous rocks is right at the level of the top of the cliff, so the blocks of sandstone fell from above. You can observe this same disconformity in creek bottoms along some of the hiking trails in the park. Boulders of quartzite, granite, and greenstone also indicate that glacial deposits exist here, but only as a thin veneer. Above the Dakota sandstone and glacial deposits and capping the hilltops are loess deposits.

From the top of the hills, you can clearly see a braided river system. Most of the time, elongate tapered sandbars are visible in the channel, sometimes associated with trees or log jams. The water immediately downstream from the logs moves more slowly, and as

Pennsylvanian rocks exposed at Schramm Park State Recreation Area.

it loses energy, it drops its load of sand, initiating the formation of a sandbar. The logs essentially anchor the sandbar. Typically, one side of the sandbar slopes smoothly into shallow water while a steep little erosional bank several inches high forms on the other side, sometimes adjacent to a channel. This asymmetry reflects depositional and erosional forces that continually change the shape, size, and position of the sandbars and intervening channels. For those who visit the river year after year, the changes are quite evident and challenge paddlers who must seek out the deeper channels to avoid ending up beached on one of the many shallow sandbars. Much of the river is only inches to several feet deep for a good part of the year. Larger, deeper channels meander back and forth. Just downstream from Schramm Park, a major channel curves into the bank, and large blocks of riprap have been placed along the slope to protect it and the road from erosion. Over the years, as dams moderated the scouring springtime flows, vegetation established itself and stabilized some of the larger sandbars, creating islands. One such island is about a quarter mile upstream from the park.

The Platte River valley narrows considerably in this area compared to its width upstream. The restriction is likely related to the exposed, resistant bedrock of Pennsylvanian and Cretaceous strata. The reason for the bedrock exposure is not clear, but it restricts limestone quarry activity to this stretch of the Platte. A particularly large operation occurs downstream in Louisville. Pleistocene gravels to the southwest near the Kansas-Nebraska state line indicate that the ancestral Platte used to flow south to here, probably along the margin of the glacial ice lobe and moraine that extended down the axis of the Missouri River. Later, a smaller drainage feeding into the Missouri River may have etched its way west until it captured and diverted the Platte River into its modern, easterly path. This theory also helps explain why the valley is more narrow here.

Finally, Schramm Park provides an opportunity to see the relationship between geologic substrate and vegetation. Note how the vegetation down on the floodplain is distinctly different from that on bluffs and hill crowns. Cottonwoods flourish on the alluvium of the floodplain where water is consistently available, but they are absent from the drier, rocky hills where junipers and scrubby bushes and trees are more common.

West of exit 432, I-80 descends into the valley of the Platte River. A rest area for the westbound lane provides a good vantage point. Looking upstream to the north, you see a broad lowland, which is the combined floodplain of the Platte and Elkhorn Rivers; the confluence is just north of here. While dams upstream control spring runoff along the Platte, ice dams sometimes form in early spring. In March 1993, ice dams caused extensive flooding that closed the interstate for a short time.

The low country to the northwest, a feature easily distinguished on topographic maps, is the lower end of the Todd Valley. The ancient Platte occupied this valley before the river shifted to its present course. Lincoln pumps its municipal water supply from the shallow aquifer of this old alluvial plain at Ashland, just upstream from the I-80 bridge over the Platte. Omaha has a well field in the alluvium of the floodplain of the Platte farther downstream and plans to build another one farther upstream.

As you drive onto the Platte River floodplain, look downstream to the east to see a valley with quite a different appearance. The

floodwaters

Platte River

View from Mahoney State Park tower looking northeast over the 1993 floodwaters backed up behind an ice dam that formed downstream of the I-80 bridge.

Platte River below this point has cut down into the Paleozoic and Mesozoic bedrock and is much more confined between steeper valley walls. You may catch glimpses of the Cretaceous Dakota sandstone in patchy outcrops along the north valley wall, especially in the winter. To the east of the interstate, you can see a small limestone quarry dug down through the Dakota sandstone into the Pennsylvanian limestones and shales below. Small quarries such as this are common from here to the mouth of the Platte, and some still operate.

South of the Platte River, I-80 quickly ascends the south valley wall. Relatively recent roadcuts around exit 426 and Mahoney State Park expose more Dakota sandstone. The interstate continues across a broadly level surface with gently rolling hills of glacial till mantled

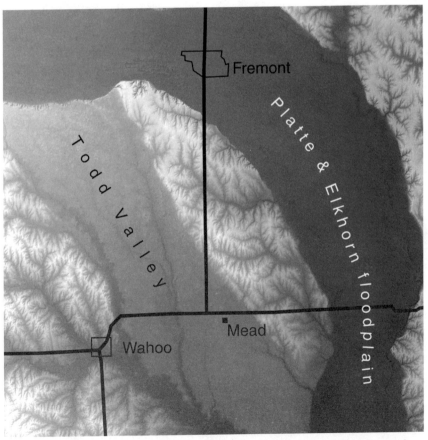

This shaded-relief map shows the abandoned Todd Valley along with the modern Platte-Elkhorn River valley. The map is made from a 30-meter digital elevation model composite of twelve U.S. Geological Survey 7.5-minute topographic quadrangles. Image area is approximately 24 miles wide, with two-fold vertical exaggeration.

with loess and dissected by small streams. The interstate roughly parallels the valley of Salt Creek, which it crosses at about milepost 407. Wind consistently sweeps through this area, and Lincoln Electric Service has erected two wind turbines visible just north of the interstate at about milepost 406. They are capable of generating 1,320 kilowatts per hour in a 33-mile-per-hour wind.

The University of Nebraska State Museum

Public museums not only educate and entertain, they also promote scientific activity that pushes forward the frontiers of knowledge. Natural history museums serve as a repository for specimens and information that is the basis for that knowledge. Housing, cataloging, and preserving specimens is difficult, expensive, and often unappreciated, even by those who make use of the knowledge in some form. Relatively few museums take on this task, and museums that house major research collections of fossils are rare treasures indeed. The University of Nebraska State Museum in Lincoln is one of these great museums, preserving specimens not just from Nebraska but from around the world.

Morrill Hall, on the University of Nebraska campus, houses the museum's displays that are open to the public. In front of Morrill Hall stands a life-size sculpture of "Archie," (short for *Archidiskodon imperator,* an obsolete taxonomic name for the imperial mammoth) a mammoth whose skeleton is in the museum collection. From the archaic proboscideans of Miocene time, to the gigantic mammoths of Pleistocene time, Nebraska has produced a tremendous collection of fossil elephants, both in quantity and quality. On entering the museum, the first thing you see is Elephant Hall, a display of elephants unequalled anywhere else in the world. Skeletons of proboscideans of all geologic ages are mounted in the large exhibit, and an impressive mural of mammoths covers the wall at one end of the hall.

Fossil mammals, for which Nebraska is rightly famous among paleontologists, dominate the museum exhibits. Halls to the left and right of the entrance display mounted skeletons of fossil mammals characteristic of the White River, Arikaree, and Ogallala sediments of Tertiary age as well as fossils of Pleistocene time. There are horses, rhinos, camels, carnivores, rodents, and other animals, including some very peculiar and unfamiliar beasts. The centerpiece of one room is a gigantic camel skeleton that towers over the visitor.

A series of interpretive displays takes the visitor through the Paleozoic and Mesozoic history of Nebraska with specimens, dioramas, and artwork, although not all of the fossils come from Nebraska. Fossil invertebrates and some fish highlight the Paleozoic display. The seas of Cretaceous time are represented by marine invertebrates such as the inoceramid clams and ammonoids, and by marine reptiles. A ceratopsian dinosaur and a tiny mammal represent the terrestrial fauna of late Cretaceous time.

On the second floor of exhibits, one hall is devoted to dinosaur fossils from the Jurassic Morrison formation. The Morrison lies deep beneath the surface in Nebraska, so these fossils were collected in

Ancient and modern elephants march down one side of Elephant Hall at the University of Nebraska State Museum in Lincoln. From left to right, they are a modern African elephant and calf, a modern Asian elephant, a dwarf mammoth from the Mediterranean island of Sicily, a Columbian mammoth from Pleistocene sediments in Nebraska (the largest mounted skeleton), and a Jefferson's mammoth from Pleistocene sediments of Nebraska. Note the extreme difference in size between the Columbian mammoth and the dwarf mammoth just in front of it.

Utah. A skeleton of an *Allosaurus* accompanies a life-size reconstruction of the beast, and a *Stegosaurus* skeleton is also on display. This upper floor also has displays of minerals. Contact the museum or visit their website for more information about resources and educational and outreach programs.

A few blocks away at Nebraska Hall is the business end of the museum. Although not open to the public, this building houses the research collections, preparation labs, and offices of paleontologists—this is where the real scientific work of the museum is done. You can contact scientists at the museum through the website. Also housed in Nebraska Hall is the Conservation and Survey Division of the University of Nebraska—the official name of the state's geological survey. The survey office is open to the public, and survey publications are available there. The University of Nebraska Department of Geosciences is situated in Bessey Hall, right next to Morrill Hall.

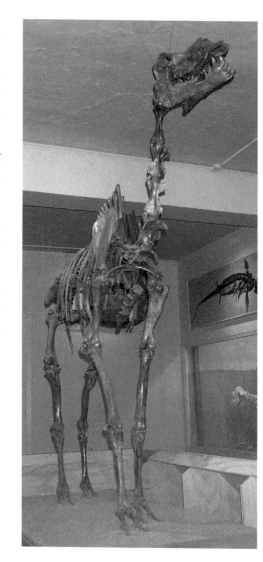

Skeleton of the camel *Gigantocamelus* from Pliocene sediments in Garden County, Nebraska, on exhibit at the University of Nebraska State Museum in Lincoln.

The ecologically distinctive Salt Creek drainage basin, which I-80 crosses as it passes through the Lincoln area, is home to 276 saline wetland bodies with an aggregate size of 5,644 acres. Because of their salty character a unique flora and fauna reside here. Best known is the endemic Salt Creek tiger beetle of which fewer than a thousand may remain. These wetlands are some of Nebraska's most threatened ecosystems. The salinity, primarily derived from groundwater flow through the underlying Dakota sandstone, is enhanced by surface evaporation. The biggest wetland is associated with Salt Lake, a remnant of which is Capitol Beach Lake just east of exit 397 on the southeast side of I-80. The wetland borders the northwest side of the interstate, which divided Salt Lake into these two parts. Additional saline wetlands and soils occur where the interstate crosses Little Salt Creek between exits 401 and 405 and where it crosses Salt Creek, which drains northeast to the Platte River.

Native Americans obtained salt from this lake and settlers attempted to mine the salt in the 1850s. In the 1860s and 1880s, they drilled deep holes in hopes of striking the subterranean motherlode layer of salt. However, no such layer was found even 2,463 feet deep. While the source of the salt is still not clear, it is likely from salt minerals distributed in small patches and pore

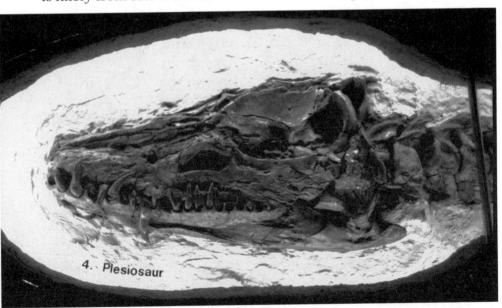

4. Plesiosaur

The skull of a plesiosaur, a marine reptile from the Cretaceous sediments of the Western Interior Seaway, on exhibit at the University of Nebraska State Museum in Lincoln.

Air photo shows the former extent of Salt Lake, including the drained portion and the remnant, Capitol Beach Lake. You can see the channelized and redirected drainage to the north and west. —*Photo from U.S. Geological Survey TerraServer website*

spaces throughout some portion of the Dakota sandstone. The discovery of abundant and more cheaply mined salt in Kansas put an end to salt mining aspirations in the Lincoln area.

With a major metropolitan area centered in the Salt Creek drainage basin, development versus conservation tensions exist. Some of the drainages constitute flood hazards that are controlled in much the same way as Big Papillion Creek in Omaha—by channelization and a series of ten dams on tributaries surrounding Lincoln. The channelization, in particular, tends to lower groundwater tables and damage or destroy adjacent saline wetlands. Many wetlands have been purposely drained. One served as a city dump for a period of time. Various entities have identified and classified the remaining wetlands and are attempting to preserve them. The Nature Conservancy controls a 200-acre site, the Little Salt Marsh Fork, north of Lincoln.

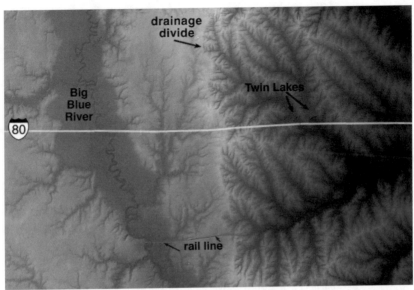

This shaded-relief map, constructed using a 30-meter digital elevation model, shows where I-80 crosses a major divide *(down the middle of the map)* between the Salt Creek drainage basin to the east and the Big Blue River drainage basin to the west. A subtle feature when you drive over it, the divide is striking in this image. This boundary is also the western margin of the youngest glacial tills in Nebraska. The glacial deposits and associated topography appear to have controlled the development of the drainage divide and pattern. The map covers the Pleasantdale and Milford 7.5-minute U.S. Geological Survey quadrangles, showing an area approximately 10 miles wide.

West of Lincoln at about milepost 384, I-80 crosses a drainage divide and drops into the valley of the Big Blue River. The Big Blue flows south along the edge of the dissected glacial surface, its path probably determined by the margin of this deposit of till. Tributary streams draining into the Big Blue from the east are very short and steep, while those to the west flow long distances across an area of low relief, their headwaters reaching almost to the Platte River. The low relief of the more eroded, older glacial tills that underlie the western area contrast with the steeper relief developed on the younger tills to the east. About 10 kilometers west of York the interstate passes beyond the western limit of the till. The ice sheets of Pleistocene time never reached farther west than this.

At about milepost 316, I-80 crosses a barely recognizable divide and descends into the Platte River valley, following it for 200 miles

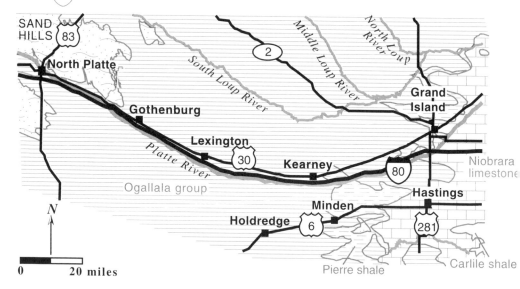

Geology along I-80 between Grand Island and North Platte. —*Modified from Dreeszen and others, 1973; Diffendal, 1991; and Burchett and Pabian, 1991*

between Grand Island and Ogallala. The valley walls slope gently and are not very high, and the wide floodplain is very flat and seemingly featureless. But, even though there is a lack of outcrops, a dynamic geologic agent is at work—the river.

The Platte River is a classic example of a braided stream. Sand and gravel bars split the flow of the river so that it is difficult to distinguish a single main channel. The many shallow, anastamosing channels, which give braided streams their name, sprawl over a wide area—it's been said that the Platte is a mile wide and an inch deep.

What makes a river become braided rather than meandering? Important factors include a relatively steep gradient, coarse sediment load, and widely fluctuating discharge. The retention of water in reservoirs for flood control and power generation, and the withdrawal and consumption of water for irrigation has greatly altered the last of these, and we may see future changes in the dynamics of the Platte River. In fact, the lower Platte, where I-80 crosses it between Omaha and Lincoln, still has some braided channels but no longer exhibits all the typical features of a braided stream. Less variation in the discharge because of upstream reservoirs has allowed vegetation to establish itself on and stabilize sandbars, so they form islands and banks that confine the channel.

Sandhill cranes in a field adjacent to the cottonwood-lined Platte River.

Take a good look at the Platte River as I-80 crosses its many channels near Grand Island. The interstate remains north of the channels until well west of Lexington; the only thing that indicates a river is there are the dense groves of cottonwoods growing alongside the channels. A consistent supply of water and the absence of human development contribute to their survival. The river at one time or another occupied all of the land along this stretch of I-80, and elongate depressions reveal the course of abandoned channels. This close to the river, the water table is shallow, and water fills many extensive low areas, forming marshes and sloughs.

This broad band of wetlands in the midst of an otherwise dry but biologically productive plains ecosystem provides indispensable habitat for migratory waterfowl. Every spring, thousands of bird-watchers flock to the stretch of the Platte between Grand Island and Kearney to see hundreds of thousands of geese, ducks, and other waterbirds, and the eagles that follow them as they head north to their breeding grounds. The sandhill cranes and rare whooping cranes usually steal the show. Huge flocks of birds have been migrating through Nebraska and stopping to rest along the Platte River at least since Pleistocene time and possibly longer. Our stewardship of the river and its wetlands may determine whether the migration will continue.

Here and there, road construction workers have excavated gravel from the floodplain sediments for use as road fill or other construction. When these shallow sand and gravel pits are dug below the water table, they quickly fill with water. Abandoned gravel pits, called borrow pits, are common along I-80. The small lakes that fill them have been developed into wayside areas such as Mormon Island State Recreation Area at mile 312. Waterfowl, such as snow geese in the fall, also use these lakes.

Fort Kearney, on the south side of the Platte, was where the Oregon-California, Mormon, and Overland Trails converged, forming the Platte River Road section of these trails. Pioneers followed the Mormon Trail from Omaha and Council Bluffs, or the Oregon Trail from St. Joseph or Independence, Missouri by way of the Rock Creek Station near the Nebraska-Kansas border. But they met at Fort Kearney and followed the Platte all the way to the Rockies, with the Mormon Trail remaining on the north side of the river and the Oregon Trail remaining on the south side. Later, the Pony Express followed the Platte River Road. The fort is now a state historical park with a museum and some restored buildings. You can reach it from exit 279 or the Kearney exit 272. Look for the high ground visible to the north, which is the valley wall, a bit more distinct and closer to the interstate here than farther east.

West of Kearney at about milepost 259, the interstate crosses the Kearney Canal, just one of several such canals that divert water from the Platte River for irrigating crops across the floodplain. The canals rejoin the river, or one of its tributaries, farther downstream. Thus, water that is not consumed by the crops returns to the river.

West of about milepost 250 the floodplain widens, and the north valley wall is more distant. However, the southern valley wall becomes progressively more distinct. Though largely obscured by the cottonwoods lining the channels, you can occasionally see the southern edge through gaps in the trees.

West of Cozad (near milepost 216), I-80 begins to cross and recross the multiple channels of the Platte, which seem to unravel a bit as we go upstream. At Gothenburg, one of the stations that served the famous but short-lived Pony Express has been restored and moved to the city park. West of Gothenburg, the valley narrows and the valley wall to the north is much closer. Looking across the valley to the north, you may be able to make out sandy sediments atop the valley wall that mark the edge of the Sand Hills. Near exit 199 you can get a good look at the south valley wall with its steep, juniper-lined ravines cutting into Ogallala sediments.

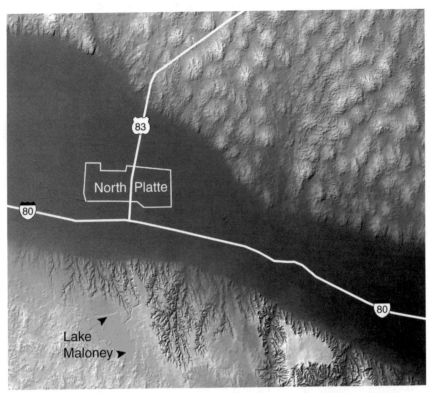

This shaded-relief map shows the constriction of the Platte River valley just east of the town of North Platte. Note the distinct difference between the north valley wall capped by near-circular domal dunes of the Sand Hills and the sharply incised Ogallala sediments of the south valley wall. The map was constructed using a 30-meter digital elevation model of six U.S. Geological Survey 7.5-minute quadrangles (Maxwell, Maxwell Northeast, Maxwell southwest, Lake Maloney, North Platte East, North Platte West). Image is approximately 17 miles wide and vertical exaggeration is two-fold.

Interstate 80 crosses the south channel of the Platte at milepost 183. The major forks of the Platte river join near here. The North Platte flows out of Wyoming, across the Panhandle of Nebraska, and around the north side of the town of North Platte to meet the South Platte, which emerges from the Rockies near Denver, flows across eastern Colorado, and passes the town of North Platte to the south. The headwaters of both branches of the Platte arise in the Colorado Rockies near the headwaters of the Colorado River.

Interstate 80 follows the south bank of the South Platte River between the towns of North Platte and Ogallala, as did the Oregon

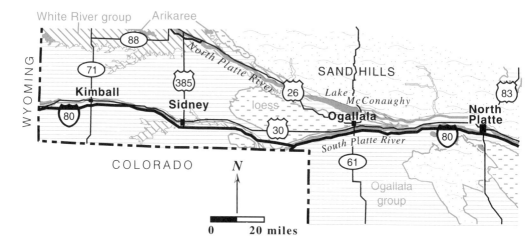

Geology along I-80 between North Platte and Wyoming. —*Modified from Diffendal, 1991; Swinehart and Diffendal, 1997; and Burchett and Pabian, 1991*

Trail. For several miles to the west of the confluence, the North and South Platte share a common floodplain, but near Sutherland a narrow ridge of Ogallala sediment separates the two drainages and widens to the west into the Cheyenne Tableland. The valley of the South Platte is quite restricted here, and at milepost 160, the river is so close to the south valley wall that the interstate must climb up above the valley floor and over a ridge called O'Fallon's Bluff, just as the Oregon Trail did. You can see wagon ruts on the bluff at the rest area for eastbound traffic. For several miles along this stretch of I-80, the Sutherland Power Station is visible to the south. This generating station makes electricity out of Wyoming coal.

The valley of the South Platte narrows where it cuts through the thick wedge of Ogallala sediment, so that at exit 145 to Paxton, you can easily see both sides of the valley. At milepost 130, ledges in the slopes to the north are actual exposures of rocks of the Ogallala group. These ledges are caused by slower erosion of the siliceous mortar beds and by ancient calcareous soil horizons that have selectively cemented the poorly consolidated sands.

Many pioneers following the Oregon Trail left the valley of the South Platte near the site of present-day Ogallala to cross over to the valley of the North Platte River. (The Mormon Trail, which had remained north of the Platte River, followed the North Platte from the confluence.)

At milepost 120 a large feedlot sprawls along the interstate. As in so many small towns in rural Nebraska, feeding cattle is important to the economy of Ogallala. Such operations are usually on the edge of the floodplain of a major river because of the need for a reliable supply of water. At the same time, a floodplain is an environmentally problematic site for concentrated animal waste. Also at milepost 120 a 1940s innovation—the center-pivot irrigation system.

At milepost 102, the interstate divides. I-76 follows the valley of the South Platte as it angles to the southwest toward Colorado and Denver. I-80 continues westward, soon climbing out of the valley of the South Platte, so that at milepost 98 it is on the level surface of the Cheyenne Tableland. It crosses the valley of Lodgepole Creek, a tributary of the South Platte, at about milepost 90. I-80 parallels the creek all the way to Wyoming, sometimes descending into the valley, sometimes ascending the valley wall to the tableland. The uppermost accumulation surface of the Ogallala formation forms the level surface of the tableland. Lodgepole Creek cut into the surface, exposing ledges of Ogallala sediments in the valley wall just east of Lodgepole at exit 76. A little farther west at milepost 63, the sediments are exposed in a small canyon. The hills are capped by sand and gravel deposited by Pliocene streams that flowed across the surface and covered by loess in places.

Good outcrops of the Ogallala sediments are just west of Sidney, and the eastbound rest area at about milepost 52 is perched on the south side of the valley in the midst of Ogallala exposures, a favorite hangout for rattlesnakes in the summer. The resistant, calcareous soil horizons that form the ledges control the development of the topography, producing flat-topped buttes with stepped slopes. Once the resistant layer has been removed, weathering and erosion quickly strip away the softer material, down to the next resistant layer. The top of the slope, protected by a resistant cap rock, develops a distinctive flat top, and the easily eroded softer material forms steep slopes punctuated by lower, resistant ledges.

At milepost 53 you can see the pumps of a small oil field. Nebraska is not a big oil-producing state, but some is present in the subsurface of the Panhandle. This part of Nebraska includes the edge of a large depositional and structural basin that lies mostly within Colorado, the Denver-Julesburg Basin. Thousands of feet of late Cretaceous marine sediments accumulated in this rapidly subsiding area. Such deep burial heated organic-rich layers enough to form petroleum, which migrated upward until trapped by impermeable layers. Most of the production is from wells that tap the Cretaceous Dakota sandstone.

Agriculture is restricted to the floodplain, close to a source of irrigation water in the narrow valley of Lodgepole Creek. Away from the creek, vegetation quickly becomes sparse. In crossing Nebraska from east to west, you gain about 4,000 feet in elevation and go from a moderately humid climate that supports fairly continuous vegetation with normal rainfall, to a semiarid climate where reliable water sources are scarce. This change takes place gradually across the state, but the contrast between the extremes is striking.

Between Dix near milepost 29 and Kimball, the valley widens and fewer outcrops are visible. But west of Kimball, I-80 passes the familiar buff cliffs of the Ogallala sediments. The town of Pine Bluffs, just inside Wyoming, is named for the pine-covered escarpment at the state line that marks the last extensive exposures of Ogallala sediments you will see along I-80 if you are heading west. Lodgepole Creek has cut through the Ogallala sediments in this westernmost part of Nebraska to expose the underlying White River sediments of the Brule formation, which can be seen at Pine Bluffs.

U.S. 6
Arapahoe—Colorado
126 miles

Between Arapahoe and Culbertson, U.S. 6 follows the Republican River valley. Just east of McCook is the Harry Strunk Memorial Rest Area, a good place to see outcrops of the Ogallala group in a long railroad cut at the base of the bluffs. Insect burrows and tunnels, burrows of larger organisms, and carbonate-rich soil horizons are well preserved here in some light green mudstones. Preserved yucca roots and hackberry seeds tell us the vegetation at the time was somewhat similar to modern plants. To have developed such widespread and distinctive soil horizons, however, rainfall must have been much more seasonal. A greater input of rain into the soil followed by arid conditions could promote greater carbonate movement and soil development. More than 10 feet of loess mantles the Ogallala group in these outcrops.

West of McCook, outcrops of the Ogallala group strata become much more common. The well-cemented soils often form small horizontal ledges that can be traced across the landscape. Some 10 miles west of McCook, you can see oil pumps south of the road. This oil field is one of several that dot this corner of Nebraska. Cretaceous sandstones hundreds of feet below the surface are the reservoir rocks that are the main targets of the wells, and production is fairly limited. A very tight cluster of small earthquakes occurred in the late 1970s in this area, as a result of oil recovery.

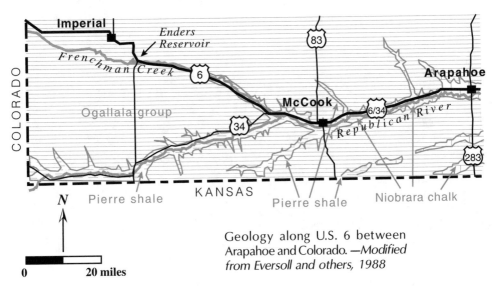

Geology along U.S. 6 between Arapahoe and Colorado. —*Modified from Eversoll and others, 1988*

A common technique used to get more oil out of the ground is to pump fluids down into the oil field to increase the pressure and drive more oil out. This increase in pressure can also trigger earthquakes if the stress conditions in the crust are right.

West of Culbertson, U.S. 6 leaves the Republican River and follows Frenchman Creek. Enders Reservoir, a good place to see the underlying Ogallala group strata, has excellent lakeshore outcrops near the campgrounds. At least four to five stacked, calcified soil horizons are present in the Ogallala sediments here, and you can see many of the burrow and root structures typical of these soils. The ancient soils are present in everything from gravel to mud layers, developing in the top part of river channel and floodplain deposits alike. To the west you can see small badlands in slopes adjacent to the road. As elsewhere, these are inactive at present, but likely go through cycles of activity and inactivity in response to climate change.

A few miles west of Enders Reservoir, small dunes to the south are outliers on the fringe of the Sand Hills. Aside from these, U.S. 6 has left Frenchman Creek and traverses a very flat surface, a remnant of the old Ogallala depositional surface that once continued gradually upwards to the tops of the Rocky Mountains. Only within

Outcrops of Ogallala group strata along the shoreline of Enders Reservoir.

the last few million years have the ancient mountains been exhumed from underneath an apron of debris that took tens of millions of years to bury them. This small remnant was once continuous with the Gangplank, the tableland between the North and South Platte Rivers that has been used by roads and railroads to ascend into Wyoming. Another small piece of the surface remains near Alliance.

U.S. 20
U.S. 275—Valentine
111 miles

Between the end of U.S. 275 and Stuart, U.S. 20 follows the upper reaches of the Elkhorn River. West of Stuart, the landscape changes noticeably as the route leaves the Elkhorn valley and travels over and among sediments of the Ogallala group, the Sand Hills, and, here and there, distinctive gravels known as the Long Pine formation. Just east of Valentine, U.S. 20 crosses the Niobrara River valley, which is incised deeply into the Ogallala group strata.

Just east of O'Neill, the valley of the Elkhorn River is more than a mile wide, but the river is little more than a creek. When a river seems too small to have created its floodplain, geologists call it *underfit* and assume it was once much larger. There are two possible explanations for why the Elkhorn River may have been larger in the past. An obvious possibility is that wetter climatic conditions in the past produced a much more vigorous river. However, the modern drainage pattern offers another explanation.

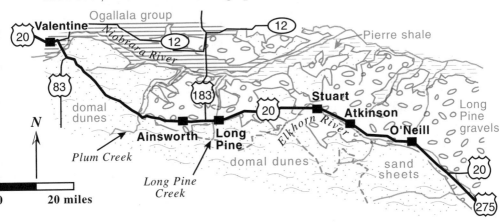

Geology along U.S. 20 between U.S. 275 and Valentine. —*Modified from Swinehart, 1989; Swinehart and others, 1994*

The drainage basin of the Elkhorn River from east of O'Neill to Newport has an interesting asymmetry. Most of the drainage basin occurs southwest of the river, and just a few miles northeast of the river is the drainage divide that separates the Elkhorn River from the Niobrara River. The Niobrara River, only about 15 miles away at one point, is deeply entrenched, while the Elkhorn sits high in the landscape. Along the Niobrara River near Valentine, late Wisconsinan (12,000 to 13,000 years old) terraces are almost 200 feet above the present river level, indicating that the Niobrara River has become entrenched within the past 12,000 years. As the Niobrara cut down, its tributaries cut back, capturing the upper drainage of the Elkhorn and, perhaps as recently as 5,000 years ago, depriving that river of some of its water. As the Niobrara River drainage continues to cut back into the landscape, it may rob more and more of the Elkhorn's drainage basin.

Just south of U.S. 20 near Atkinson, there are sand and gravel pits in the Elkhorn alluvium. The source of the gravel is likely material winnowed from other, older channel deposits in the area as the present cycle of erosion eats through the accumulated Cenozoic sediments. You can also see small dune forms on the floodplain. Some 10 miles west of Atkinson, U.S. 20 leaves the Elkhorn valley and travels on a relatively flat surface known as the Ainsworth Table. Patches of dunes can be seen to the south.

Long Pine Creek cuts down through the Ainsworth table and provides some excellent exposures—the type section—of the Long Pine sands and gravels, especially on the east side of the creek. These sands and gravels are part of a sheetlike deposit more than 100 feet thick in places. The igneous and metamorphic clasts in the gravel came from the Rockies. A braided river cut channels into the underlying Ash Hollow formation of Miocene time and deposited these gravels. Some geologists suggest that a major drainage system, a Platte River ancestor, crossed the state with an east-northeast orientation and deposited the Long Pine gravels and the similar Broadwater formation deposits west of Lake McConaughy. Horse and other fossils in the Long Pine gravels indicate they were deposited between 3 and 2.5 million years ago in late Pliocene time. Groundwater feeds Long Pine Creek to a significant degree, as is evident from the lack of supporting drainage. You can find a unique flora that is a relict of the Ice Age in the creek canyon at Long Pine State Recreation Area. As in the Niobrara River valley, the cool canyon enables plants that prefer cool climates to survive much farther south than their normal range.

At Ainsworth municipal airport and miles east of town, you can see gravel pits from the road. Elsewhere along this section, low-relief, whitish ledges typical of the Ash Hollow formation surface in places where the Long Pine gravels are missing. Just west of a section of the Sand Hills, U.S. 20 crosses Plum Creek, which may represent some of the drainage the Niobrara River captured from the Elkhorn. Wetlands, meadows, and small lakes are present in the interdune depressions. Several creeks connect to the concentration of lakes and wetlands in the Valentine National Wildlife Refuge, and this area may have been part of the Plum Creek drainage before the sand dune invasion. Widely spaced barchan and domal-ridge dunes are dominant, with smaller linear dunes and sand sheets on the periphery.

Where U.S. 20 crosses the east side of the Niobrara River valley, it descends from the tableland, through the ledge-forming cap rock member of the Ash Hollow formation, and into the sands of the Valentine formation. Near the railroad bridge to the north, quarries in the Valentine formation have yielded fossil mammals of mid-Miocene time. The river flows on the silt-rich Rosebud formation. On the west side of the road, patches of dune sand mantle the terrain.

U.S. 20
Valentine—Wyoming
197 miles

U.S. 20 between Valentine and the Wyoming border near Harrison crosses a good sampling of the Cenozoic rocks and sediments that characterize so much of the geology of western Nebraska. But you will not encounter them in a simple spatial or chronological order. Although these sediments are relatively flat lying—little affected by later disturbances—their deposition by wind and stream was not uniform to start with. Episodes of erosion between depositional intervals, including the erosion that carved the present topography, have conspired to create a complex pattern of discontinuous sedimentary layers.

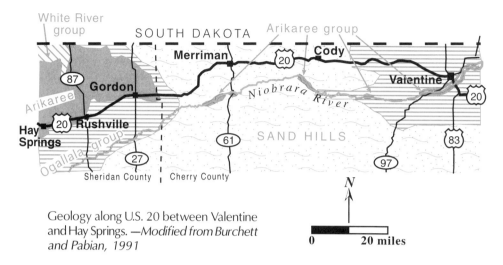

Geology along U.S. 20 between Valentine and Hay Springs. —*Modified from Burchett and Pabian, 1991*

0   20 miles

Valentine sits near the confluence of the Niobrara River and Minnechaduza Creek. A prominent escarpment extends north of the road, along Minnechaduza Creek, and eastward along the north side of the Niobrara River valley. Exposed in the escarpment are the lower units of the Miocene Ogallala group, a major component of the High Plains aquifer. In some places the Ogallala sediments are undivided, but here the lower slope exposes the Valentine formation and the overlying resistant ledge is the Cap Rock member at the base of the Ash Hollow formation. The Cap Rock member maintains the escarpment by protecting the softer sediment below from erosion. The entire section is well exposed a short distance east of Valentine along the Niobrara River where Nebraska 12 climbs out of the valley. U.S. 20 follows the broad and fairly straight valley of Minnechaduza Creek west of the town.

From Valentine west to Nenzel there are small exposures of Ogallala sediments, mostly Ash Hollow formation. Locally, fine, loose sand occurs in patches on top of the Tertiary rocks, and just west of Crookston the sand bodies form dunes, once active and now stabilized by a grass cover. This is the northern margin of the Sand Hills.

South of the road, the Niobrara River has cut a canyon—steep by Nebraska standards—down to the Tertiary rocks, dividing the Sand Hills into a smaller northern and a larger southern section. U.S. 20 crosses through almost 60 miles of the northern section between Nenzel and just east of Gordon. Near Cody and westward,

Air photo of Sand Hills between Eli and Merriman along U.S. 20, about 2 miles of which stretch across the photo center. The dark areas are flat interdune depressions with groundwater-fed meadows and denser, greener vegetation. A north-northwest/south-southeast linearity to the sand dunes represents dune elongation in the direction the wind blew. These are smaller and younger linear dunes superimposed on older, degraded, larger barchan dunes.
—*Photo courtesy of U.S. Geological Survey*

the landscape is characteristic of a great sand sea. At present, a mixture of grasses, forbs, and herbs covers and stabilizes the dunes so that they no longer migrate with the wind as they once did. But the original dune topography is little altered from a time when the sand was actively moving throughout the region. The large hills surrounding the road have steep south- and southeast-facing slopes that were the downwind sides of the dunes. The gentler, upwind slopes typically face north and northwest. The dune forms are complex here, and you can readily see blowout scars.

Cottonwood Lake State Recreation Area, just east of Merriman, is a good example of a groundwater-fed lake and represents an environmental microcosm in the Sand Hills. The lake traps pollen and windblown sediments, preserving a record of local changes within the bottom sediment.

Between Merriman and Gordon, U.S. 20 crosses the western border of the Sand Hills and descends to a flat area with irrigated fields. You can see some small, table-top mounds from the road. White resistant units, calcified ancient soil horizons within the Ogallala sediments, cap the mounds. These soils formed in semi-arid climates, not greatly different from present conditions. From Gordon to Crawford, the road crosses into progressively older rocks.

East of Chadron, U.S. 20 descends from the relatively flat plains to the east through a series of pine-covered ridges and draws into low, rolling hills. A prominent topographic ridge extends fairly continuously from Chadron west to the Wyoming border and is easily visible on satellite images. Ponderosa pines grow preferentially on the crest and slopes of the ridge, a characteristic that gives it the name Pine Ridge. Sediments of the Arikaree group, deposited during late Oligocene and early Miocene time about 28 to 19 million years ago, form the ridge. These homogenous, light tan, fine-grained sandstones and siltstones are riddled in places by numerous discontinuous layers of large, elongate and bulbous, carbonate concretions that grew after the sediments were deposited. Small variations in the chemistry of the groundwater, from which minerals precipitated, cemented the sediments more heavily in some areas than others. The homogenous character, abundant concretions, and a greater degree of cementation and sand make

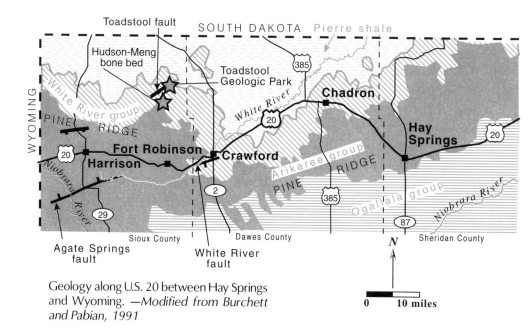

Geology along U.S. 20 between Hay Springs and Wyoming. —*Modified from Burchett and Pabian, 1991*

the sediments of Pine Ridge more resistant to erosion than overlying and underlying sediments.

An indentation in the ridge gives rise to the headwaters of the
White River. The Toadstool fault along the northern edge of Pine
Ridge and faults along the White River may have helped control
the large-scale topographic pattern of the ridge and valley. The
faults are part of the Colorado-Wyoming lineament and may be
related to a reactivated Precambrian suture deep in the basement
rocks below.

Near the base of Pine Ridge is the contact between the more
impermeable, volcanic-rich Brule formation of the White River
group (40 to 80 percent volcanic ash) and somewhat sandier sediments of the overlying Arikaree group (20 to 60 percent volcanic
ash). This creates local, perched aquifers and springs within the
Arikaree sediments, such as the one at the Hudson-Meng bone
bed. Water seeping down through the sediments is less able to pass
through the impermeable Brule formation, except along fractures.

About 11 miles east of Chadron, you can see a quarry to the
south of the road. It contains coarse alluvial conglomerates—its
pebbles eroded from nearby sources. These are deposits from a
younger cut-and-fill cycle, possibly Pleistocene in age. The present
valley may thus be partly a modern reexcavation of an older valley
cut into the bluffs of Arikaree sediments.

West of Chadron, the road descends farther through the stratigraphic section into the White River group. The lower part of this
unit has a more varied appearance than the overlying Arikaree,
with tan channel sands, nodular sandstones, multicolored siltstone
and mudstone horizons, and distinctive white ash layers. At the
foot of small buttes and mesas capped by resistant layers, small
badlands develop just as they have done in strata of similar age in
the Big Badlands of South Dakota. Five to six miles west of Chadron,
you can see the relatively isolated Trunk Butte to the south of the
road. The capping resistant rocks of the bluff are limestones deposited in a lake and underlain by lake silts.

To the north of the road, erosion has removed the Tertiary
sediments, exposing the underlying Cretaceous Pierre shale. Progressively older rock units are exposed from here to the Black Hills
of South Dakota. You can see the hills in the distance to the northwest from the Crawford area on an exceptionally clear day. The

Black Hills are a domal uplift elongated north to south. The south-ernmost nose of the uplift, a very subtle structure, extends into the northwest corner of Nebraska, where it gives the rocks their gentle regional dip to the south.

The contact between the White River group of Tertiary age and the Pierre shale of Cretaceous age is a major disconformity, rep-resenting a gap in the stratigraphic record of tens of millions of years. Before the basal White River sandstones, conglomerates, and ash were deposited, the normally black Pierre shale was weath-ered at the land surface and oxidized to bright yellows that you can see, for example, around the base of Trunk Butte. This ancient soil formed during the long exposure of the shale to the atmosphere before deposition of the White River group. The red exposures above the yellow soils are another soil that developed in the Chamber-lain Pass formation at the base of the White River group.

Just southeast of Crawford, uranium is mined from sandstone and conglomerate at the bottom of the White River group. This basal unit has been recently named the Chamberlain Pass forma-tion but is known in earlier work as the Chadron conglomerate. The uranium occurs as a precipitate within the pore space of the sandstones and conglomerates. It has been mined belowground since 1989 by a solution method. Hot water charged with hydro-gen peroxide and other oxidants is pumped down into the ore-bearing strata, where it leaches the uranium-bearing minerals. The leachate is pumped out and processed on-site, and only a few buildings and wellheads mark the mine. Responsibility for moni-toring the mining activity and preventing mining accidents or groundwater contamination falls largely to the mine operators and the Nebraska Department of Environmental Quality.

The action of a natural underground chemical filter formed this uranium deposit. Relatively shallow, oxygen-rich groundwa-ter tends to leach uranium compounds from the sediments it passes through. Both volcanic ash and black shales are potentially richer in uranium than many other rocks or sediments. As groundwater charged with dissolved uranium compounds traveled through the basal sandstones and conglomerates, it encountered oxygen-poor chemical conditions that caused the uranium compounds to pre-cipitate in the void spaces in the conglomerate. The mining method essentially reverses this chemical process.

Toadstool Geologic Park

About 16 miles northwest of Crawford is Toadstool Geologic Park in the Oglala National Grassland, where the White River group is exposed in badlands. Head north from U.S. 20 on Nebraska 2 and follow signs to the park. Dirt roads in this area become very slick when it rains because of the smectite clays within the basal White River group sediments. These clays form by the weathering of volcanic ash, an abundant component of the original sediment of the White River group. Some of the silt was originally composed of up to 90 percent volcanic glass shards. Luckily, the road into the park has been substantially improved with gravel.

A small sod house open to the public at the park campground gives an insight into living conditions of the early Nebraska pioneers. Water is sometimes available from a hand pump at the campgrounds, but the distinctive and strong alkali taste is unpleasant. In addition, groundwater from the Chamberlain Pass aquifer has high levels of uranium in it in places. It is best to bring your own water.

It is the water, more than anything else, that gives us the term *badlands*. To the early trappers and explorers who trekked across the spectacularly eroded White River sediments just north of here in South Dakota, the land was bad because it was desolate, hot, and dry, and what water there was to tempt the thirsty was usually loaded with alkaline salts. Anyone unfortunate enough to drink the water would discover the laxative qualities of the alkaline solution, an effect that could be fatal in such a hostile environment. Nowadays, geologists use the term *badlands* to refer to any area where intense erosion has carved relatively soft sedimentary rocks into complicated, bare exposures.

The bulk of the intricate badlands in the park are within the Orella member of the Brule formation. These multicolored strata contain a multiplicity of old river channels. An ongoing rearrangement of the stratigraphic nomenclature may reassign the lower portion of these channel sandstones into the Chadron formation. Some channels are obvious because they are filled with coarser sands. Others, however, are filled with mud. If you carefully trace layers laterally, you can find layers of fine-grained sediment truncating other layers in a low-angle contact. It is likely that windblown or finer-grained flood deposits filled in a secondary or abandoned channel that was dry most of the time. Distinctive reddish brown horizons are probably indicators of ancient soils.

A loop trail from the parking area ascends to the top of and then follows a particularly large ancient river channel. These deposits consist of crossbedded, light tan, coarse sandstones. You can see the south side of the channel wall where it cuts down into well-bedded, reddish brown siltstones and sandstones; the old channel axis is oriented roughly east-west here. In the tops of these sandstones, tracks of birds and large vertebrates are preserved. Riparian corridors attract wildlife, which can leave their imprints.

To the north, an east-northeast-trending fault called the Toadstool fault cuts the channel deposits. The fault is silicified with a fine-grained form of quartz known as chalcedony and contains some calcite veins, which often have a polished surface with small linear grooves or striae. Scoring the surface during faulting formed the

Badlands topography in the Orella member of the Brule formation at Toadstool Geologic Park. Some of the darker bands (reddish) are likely ancient soil horizons.

View looking west at Toadstool fault. The fault is located between the two arrows, with the better-bedded Orella member strata to the left of the fault, and the more massive Chadron formation strata to the right. Also note the channel sandstones that form ledges to the left in the upper portion of the Orella member.

grooves, which indicate that movement was down the direction the fault is inclined. If you walk around the trail loop, you cross the fault very close to where the trail comes down to and parallels the modern wash. The continuation of the same channel sands you have been walking on now forms the mesa top of the ridge on the north side of the fault. The north side moved up relative to the south side by about 60 feet. These types of faults are known as normal faults and are associated with crustal extension. You can easily spot other faults in the excellent badlands exposures. They have offsets from as little as less than a foot to tens of feet. Several sets of faults exist—each set defined by their common orientation. Toadstool Geologic Park is centered at the intersection of two of these sets. Just south of the Toadstool fault is another smaller, parallel fault that cuts the channel sandstones. Between the two faults is a downdropped block, known as a graben, that the trail follows while it is on the channel sandstones. Numerous sets of chalcedony veins fill fractures without any indication of offset that cut the sediments in places.

Since the fault cuts 30-million-year-old Oligocene strata, they must be even younger. It is unclear what tectonic forces could have caused faulting during that time or since. The Rocky Mountain upheaval that produced the Laramie Mountains to the west in Wyoming and the Black Hills to the north was long since complete by that time. The faults could be related in time to the formation of the Basin and Range by crustal extension, but the Basin and Range province in the southwestern United States is a long way from here. The faults could also be related to reactivation of older faults in the ancient basement rocks deep below. The north-northwest-trending set of faults could represent reactivation of deeper structures associated with the southern nose of the Black Hills uplift. The more northeast-trending set may be due to reactivation of old Precambrian suture faults.

The light tan, greenish, and whitish thick-bedded strata at the very base of the badlands to the north of the fault are siltstones of the Chadron formation. Small sand lenses deposited in river channels snake through the Chadron siltstones. Fossil mammals have been found in these strata, including the remains of brontotheres, large rhinoceros-like animals that roamed the land 30 million years ago.

The clays derived from the weathering of the volcanic ash in the finer-grained sediments of the White River group shrink and swell during drying and wetting cycles. This weathering produces the distinctive surface texture known informally to geologists as popcorn. Cracked and clotted knots of hardened clay are held only loosely to the slope, but if you dig down several inches, you will come to the more compact, unweathered, original sediment. This crumbly material actually poses a hazard to those climbing about, as it can be like walking on marbles. It also aids in badland formation as the swelling inhibits seepage of water into the ground, in turn increasing surface runoff, which in turn carves the intricate badland topography. After a thunderstorm you can dig just an inch or two into the weathered surface, and the material underneath will be dry.

The White River group preserves a great variety of vertebrate fossils of Eocene to Oligocene age, about 35 to 25 million years ago. The Orella member is the most fossiliferous part of the section. Land tortoise shells 1 to 2 feet long are abundant. Shell fragments are extremely common at the surface. After tortoises, the most common vertebrate fossils are the remains of oreodonts, even-toed hoofed mammals that were about the size of sheep but only distantly related to them. Other herbivores included small deer, giant pigs, an early camel, three-toed horses (*Mesohippus*) the size of a German shepherd, early relatives of tapirs and rhinoceroses, several mouse- to squirrel-size rodents, and rather ordinary-looking rabbits. Among the carnivores were small dogs, saber-toothed cats, and medium to large predators such as *Hyaenodon*, that are only distantly related to any modern carnivores. In the Whitney member above the Orella channel sandstone, you can find snail shells and the seeds of hackberry trees in some horizons. Early members of many groups that are common today populated the land, although most were quite different from their modern relatives. Mixed with them were creatures completely unfamiliar to us because they represent evolutionary lineages that became extinct.

Although these rocks are exceptionally fossiliferous, fossils are not as common as they once were because of heavy collecting by amateur and professional paleontologists. Indiscriminate collecting of these fossils destroys much of their scientific value and diminishes the experience of subsequent visitors who will not be able to view the fossils in place. Fossil collecting has been illegal here since 1986, and there has been one conviction for illegal collecting from here. We urge you to look for fossils but to leave your finds in place. Let photographs be your souvenir. You can see prepared and restored skeletons of many of the organisms found at Toadstool at the University of Nebraska Trailside Museum at Fort Robinson

State Park just west of Crawford and at the University of Nebraska State Museum in Lincoln.

In 1895, N. H. Darton, a U.S. Geological Survey scientist, arrived here on horseback and found a picnic table, a buckboard track, and a sign saying, "Toadstool Park." The park obtained its name from peculiar erosional features that form where a layer of jointed sandstone overlies a clay-rich layer. Wind and water erosion selectively attacks the joints in the sandstone and removes the soft, underlying clay. Rounded blocks of the more resistant sandstone perch atop pedestals of clay that they protect from erosion. They resemble toadstools and you can see some just north of the marked trail loop. The well-developed ones to which the park owes its name are off the beaten track. The erosion of the complex Orella stratigraphy gives the Toadstool badlands an intricacy of form not present in more homogenous strata, such as the overlying Whitney member of the Brule formation, which you can see at the base of Roundtop, a distinctive conical peak to the south.

A careful inspection of air photos and topographic maps reveals some preferred directions of ridge and channel orientation where there are no faults. Note for example how some of the ridges at the eastern margin extend in an east-northeast direction. This likely reflects fracture control of the erosional process. Regularly oriented and relatively planar fractures, or joint sets, occur in these sediments, although surface weathering and shrink-swell behavior of the clays often obscures these features. You can see fractures better in the sandstones and in some fresh cutbank exposures.

Part of the loop trail returns along a wash that provides insight into how badlands are formed. Although the streambed is dry most of the time, infrequent but intense rainfalls erode the badlands and wash debris into the stream. If a stream has a large enough drainage basin, flash floods can fill the channel. Undercutting of the banks by the stream and subsequent collapse of the valley walls during these events is an important mechanism of erosion. With each powerful erosional event, these badlands cut farther back into the mesas, in a process termed *headward erosion*. In this way, badlands grow and migrate with time, etching their way back into a table-land. Mesas and buttes form as the expanding badlands isolate remnants of the tableland.

Toadstool erosional forms about 2.5 feet tall in the interior of the badlands.

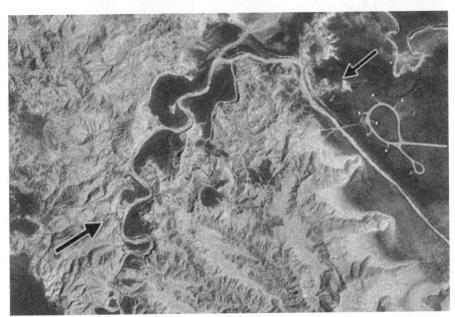

Air photo of Toadstool Geologic Park, with parking and campground to center right. The Toadstool fault extends along and between the arrows. The trail departs the campground, heads southwest, intersects the wash, and follows it downstream, looping to return to the campground. The tableland that the badlands are cut into is the flat, grassy dark area in the extreme lower left. —*Photo courtesy of U.S. Geological Survey*

Hudson-Meng Bone Bed

A turnoff from the gravel road that goes to Toadstool Geologic Park takes you up into the overlooking hills and bluffs to another notable fossil locality in Nebraska—the Hudson-Meng bone bed, which contains the densely packed bones of more than six hundred bison. Signs mark the way. Here you can learn about and participate in an ongoing scientific debate as to what caused this unique deposit. The locality is only about 3 miles walking distance from Toadstool along a trail that links the two sites and is in a draw near the distinctive hill known as Roundtop. A building has been erected on top of the site to aid in its preservation and study. In summer, students and professors actively uncover, map, and study the deposit. There is a modest entry fee.

The bone bed is named after two ranchers who discovered the site while they were excavating a pond. Thanks to their interest in natural history, they recognized its significance and brought it to the attention of scientists. Larry Agenbroad, then at Chadron State College, worked on the site for some nine years, mainly in the 1970s. He proposed that it was a bison kill site where Native Americans butchered some of the herd. The discovery of projectile points indicative of the Alberta culture of 8,000 to 10,000 years ago, and the absence of skulls, which may have been removed during butchering, supports this interpretation. However, subsequent and ongoing work by archeologists from Colorado State University and the University of Wyoming questions that interpretation.

The bone bed—the largest bison bone bed in North America—consists of the remains of *Bison antiquus,* an extinct species that was larger than the modern bison. Most of the bison are cows or calves. The bones are in light-colored, relatively fine-grained and homogenous sediment—windblown loess with some fluvial deposits. Radiocarbon dates on charcoal from the site indicate that the bone bed is nearly 10,000 years old. These sediments represent a relatively thin surface deposit draped over the eroded and much older Tertiary sediments.

View of a portion of the Hudson-Meng bison bone bed within the site building. Note that the vertebrae, ribs, and limb bones from an individual are still together. Most of at least three bison are evident in this photo.

A spring at the site emerges at the contact of the fine-grained, relatively impermeable upper portion of the Brule formation with the more permeable, overlying Arikaree group. The ranchers chose this site for a pond because of the spring. Springs occur at this stratigraphic level at many points in the Pine Ridge area. Evidence indicates the spring was here at the time the bison died.

Scientists generally agree that the bison were all killed together in a singular event. But was it by humans or by some natural catastrophe? Recent work suggest the animals died in a defensive herd position. In places, the bones are articulated, still in contact with one another at the joints, just as they were in the living animal, while in other places they have been separated and moved. The bony cores that were attached to the skulls and supported the horns are missing or not preserved, as are much of the skulls themselves. In some places, bones are intact, but in other places they are broken. If the animals were butchered at the kill site, the skulls may have been removed along with the hides. But skulls and horn cores may be particularly susceptible to the decay, scavenging, and trampling that would follow a natural disaster.

The association of human artifacts with the bones is the strongest evidence in support of the original interpretation of this as a kill site. But recent work suggests that the artifacts actually occur on a surface some 8 to 10 centimeters above the bone bed. Some artifacts may have been trampled or moved by soil-forming processes into a deeper position and thus into an apparent association with the bison bones. The spring, which may have attracted the bison herd to the site, may have later attracted people who left behind the projectile points.

What killed so many bison at one time if this is not a kill and butchering site? Alternate hypotheses include a lightning strike of closely grouped bison standing in spring waters, a grassland fire, a violent hailstorm, or a blizzard. More challenging than coming up with a hypothesis is coming up with a way to test it. Ongoing study of this site and other bison kill sites may shed light on what caused the extinction of many large Ice Age mammals in North America at this time. Scientists may have a strong case that the hunting practices of the earliest humans in North America drove these species to extinction, but they continue to vigorously debate this theory.

Just west of Crawford, U.S. 20 crosses the White River, where you can see at least two stream terraces—former floodplains. While it is hardly a river here—it can be easily jumped—in 1991 a devastating flood caused several million dollars worth of damage in the immediate area and killed one person. The river flooded the local golf course and covered the fairways with sand. The railroad was also damaged. Along this transportation artery, coal trains from Wyoming feed pulses of energy to mid-America's urban and industrial infrastructure. Each car of coal feeds a medium-size power plant for about one hour.

Looking west along railroad tracks after the 1991 flood along the White River. Floodwaters undermined and bent the tracks just beyond the crossing.

West of Crawford, U.S. 20 ascends through the Arikaree sediments as it parallels the northern slopes of the White River valley. The valley parallels a fault that is most evident in subsurface drill hole data. A small fault in a railway cut just outside of Crawford may be associated with this larger valley fault.

Fort Robinson and the Trailside Museum just west of Crawford make an interesting stop. Local fossils, including the skeleton of a mammoth, are on exhibit at the museum. The bluffs that face Fort Robinson consist of the same stratigraphic units that make up Pine Ridge near Chadron—the massive, light tan, concretion-rich siltstones of the Monroe Creek and Harrison formations of the Arikaree group. A small side road to the north of U.S. 20 winds through Smiley canyon, which cuts through these strata and provides much better exposures. Elongate and smooth pipelike concretions, some several feet long, occur in these outcrops. Similar to those at Scotts Bluff, they have a preferred orientation. The origin of this orientation is basically unknown but may be related to the direction of groundwater flow at the time they formed.

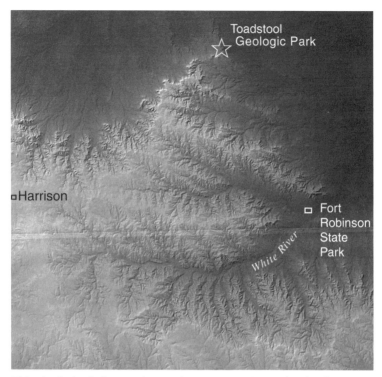

This shaded-relief map shows the western Pine Ridge landscape in Nebraska. The topography is controlled by rocks of Tertiary age, with the Harrison formation of the Arikaree group underlying the lighter, flatter, high areas and the White River group strata underlying the darker, lower areas. Image constructed from digital elevation model for twelve U.S. Geological Survey 7.5-minute quadrangles (from Bodarc in the northwest corner to Dead Mans Creek in the southeast corner).

Edge of burn along Smiley Canyon Road just west of Fort Robinson, with unburned vegetation and stabilized gully at right and outcrop patch at left.

Bluffs of Arikaree strata overlooking the road just east of Fort Robinson.

In the 1990s, a large fire just north of Fort Robinson destroyed a swath of the pines many miles long. You can still see the scars of this disaster to the north of U.S. 20. Early photos of the area, before extensive fire control, show only a fraction of the pines that cover the ridge today—the natural vegetative cover was dominantly grasses. Fires maintain grasslands by preventing sapling growth. Today, pines are planted on burned areas. Fire may have also played an important role in past environments in the high plains. Theoretically, erosion can attack the devegetated area left in the wake of a large intense fire and create a fire-induced sedimentation event in the local drainage. This theory, however, has been hard to document in sedimentary deposits.

Several miles west of Fort Robinson, U.S. 20 climbs to a relatively flat grassland plateau, on which it continues through Harrison and on to the Wyoming border. This tableland tops the Pine Ridge escarpment. The road follows this tableland into Wyoming, and you can see small outcrops of the Harrison formation of the Arikaree strata in roadcuts and as ledges.

U.S. 26
Ogallala—Wyoming
149 miles

U.S. 26 between Ogallala and Wyoming is a scenic route, with an array of interesting, well-exposed geology. West of Ogallala, U.S. 26 quickly climbs onto the Gangplank surface, remaining on it for the length of Lake McConaughy, then dropping back down to the North Platte River valley and following it for the remainder of the route. Along the way, U.S. 26 passes many famous landmarks of the Oregon Trail—Courthouse Rock, Jail Rock, Chimney Rock, and Scotts Bluff. These bluffs, buttes, and chimneys stand out from Wildcat Ridge, which parallels the North Platte. Nebraska 92 runs concurrent with or parallel to U.S. 26 for much of the way and, in places, offers closer views of bluffs and other landforms.

Just north of Ogallala, Nebraska 61 takes off to the east toward Kingsley Dam at the east end of Lake McConaughy. This short diversion from U.S. 26 is worth the trip. Roadcuts just south of the dam near the headquarters for Lake McConaughy State Recreation Area reveal excellent exposures of the Ogallala formation with well-developed soil horizons forming extensive ledges. Lake McConaughy, formed by a dam across the North Platte, is the largest body of surface water in Nebraska. The reservoir provides water for irrigation, flood control, power generation, and recreation. The extensive sandy beaches developed from erosion along the shoreline of the sandy sediments that surround the lake—the Sand Hills along the north side and the Ogallala group along the south side.

Kingsley Dam and Lake McConaughy also provide insight into some environmental concerns associated with dams. At the base of the dam, water is released in a fountain of water that sprays tens of feet in the air. You may sometimes notice a strong rotten egg smell there. Lakes often become stratified as the sun heats surface waters, which become less dense than the colder, deep water. The surface waters obtain plenty of oxygen from the atmosphere, but decomposition of organic matter in the deep water uses up oxygen. In the oxygen-poor deep waters, bacteria obtain oxygen by breaking down sulfur compounds dissolved in the water. The chemical reaction produces hydrogen sulfide, the source of the rotten egg smell. The fountain aerates the stale water released from the depths of the reservoir and puts oxygen back into it so that water quality is maintained downstream, minimizing the effect on aquatic life.

At the base of the dam you can also see some green areas where water is seeping out and flowing down to the river. The height of the water behind the dam creates tremendous pressure that drives

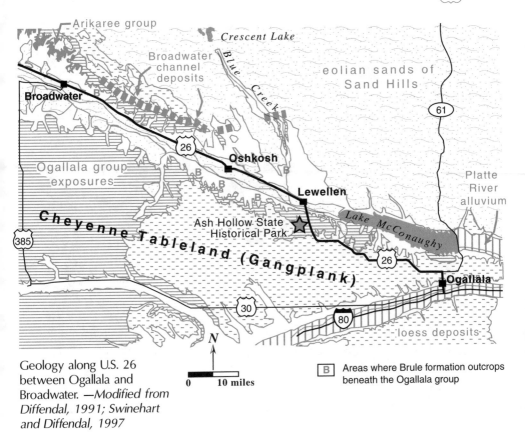

Geology along U.S. 26 between Ogallala and Broadwater. —Modified from Diffendal, 1991; Swinehart and Diffendal, 1997

0   10 miles

N

B   Areas where Brule formation outcrops beneath the Ogallala group

water through earthen dams. All earthen dams "leak" to some degree. An earthen dam must be engineered so that water flows through the structure without weakening it.

To prevent erosion, large pieces of rock and large concrete "jacks" heavily armor the reservoir side of the dam. Winds from the west sweep the length of Lake McConaughy, building large waves that have the potential for serious erosion. More than 800 feet of shoreline retreat has occurred in one area. A delta has also built out over two and a half miles at the western end.

The level of the reservoir fluctuates by tens of feet from season to season, and after it drops, you can see water seeping out of exposed shoreline sediments. Conversely, when lake levels rise, water seeps into the banks and recharges the local groundwater table, which has been elevated by the presence of the dam about 60 feet in some areas. A canal associated with a much smaller dam just downstream from Kingsley Dam diverts a substantial part of

In this satellite image, you can see the dunes of the Sand Hills to the north of Lake McConaughy with light spots indicating sand exposures, primarily in blowouts. The long, narrow strip of cultivated land, the patchwork of fields between Lake McConaughy and the South Platte River, is the Gangplank surface, which is mantled by loess from the Wisconsinan ice age. Circular areas in lower right are center-pivot irrigation systems. —*Image modified from NASA's Earth from Space website*

the North Platte River for irrigation. In a variety of legal and political forums, debate continues as to how much water should go to irrigation, how much should go to maintaining wildlife habitat along the Platte downstream, and how much should go to maintaining water levels for recreation.

Following U.S. 26 west of Ogallala takes you up onto the Gangplank surface. The Gangplank is a high, smooth, and very gently sloping surface between the South and North Platte Rivers, and the portion of U.S. 26 just south of Lake McConaughy is one of the best places to see it. Here, the surface is also known as the Cheyenne Tableland. U.S. 26 climbs up to the Gangplank from the South

Platte River floodplain near Ogallala gulch, which is cut into the sediments that form the surface. Ogallala gulch may have served as a spillway for Lake Diffendal some 10,000 years ago, so that when water levels were high enough, water flowed over the divide and into the South Platte River.

Just west of the road to Lake View campground, U.S. 26 reaches a remarkably flat area. Here the Gangplank surface is especially well preserved and is more easily noticed. All the modern topography associated with the Platte River drainage in this area has been cut into this surface since it formed. A few million years ago, and especially after about 600,000 years ago, the overall sedimentary dynamics of western Nebraska changed from accumulation of sediment to an erosion-dominated system. Using the age of volcanic ashes in terrace deposits, David Dethier estimated the South Platte River cut down into the landscape in western Nebraska at a rate of 5 to 10 centimeters per 1,000 years. This is part of a much larger story of the rebirth of the Rocky Mountains in which erosion is stripping away the apron of debris that largely covered them.

Loess mantles portions of the Gangplank. While typically only tens of feet thick, in this area the loess is up to 200 feet thick. The long-term stability of the Gangplank surface allowed a grassland cover to capture the windblown silt. The greater thickness of loess cover in this area than to the south may be the result of the proximity to an early North Platte River floodplain, which would have contributed more windblown silt. You can see outcrops of the underlying Ogallala formation, with distinctive ash horizons, along Eagle Canyon Road, which descends to Lake McConaughy to the north.

Ash Hollow State Historical Park

Ash Hollow is a small north-south drainage that empties into the North Platte River near the western end of Lake McConaughy. In Ash Hollow State Historical Park, you can see deep wagon wheel ruts that mark the Oregon Trail at Windlass Hill, evidence of the westward movement of pioneers. Wagon trains following the south side of the Platte River eventually crossed the South Platte and climbed onto the Gangplank to reach the North Platte. The trail descends from the Gangplank through Ash Hollow to the welcoming ribbon of riparian vegetation and the sustenance of the North

Platte River. Windlass Hill was the steepest part of the descent, and easing loaded wagons down this stretch required caution and considerable nerve. The wooded area and a spring in the valley below made a welcome resting spot that was often mentioned in pioneer diaries. Not a few pioneers were also buried here, many having succumbed to illness.

In the bluffs overlooking Ash Hollow and the North Platte River is a rock shelter that Native Americans occupied for thousands of years. Layers of debris deposited on the rock floor record seven different times of extended occupation—an unusually long archeological record for this area. Work from the 1940s describes these occupations as spanning a period from roughly A.D. 1 to 1700. Work in the 1980s found sites possibly as old as 9000 years in the Ash Hollow area. Further work on the rock shelter could document an even longer record of occupation.

This desirable location had the rich resources of the Platte River valley below and a strategic view that enabled the occupants to track the movements of game and observe the approach of other humans. We encourage you to visit the state historical interpretative center and the protected and enclosed rock shelter that is open to the public from Memorial Day to Labor Day.

The streambed of Ash Creek, which is usually dry, contains gravels that include pebbles of granite and metamorphic rock. A source for these pebbles is not immediately obvious. The Ash Hollow drainage does not reach any mountains with such material exposed, and the fine sediments that the stream cuts through don't contain such rocks. But, the drainage does dissect the Gangplank, which in places has a thin veneer of Pliocene or Pleistocene gravels with pebbles of this composition. This reworked material was transported from the mountains up to several million years ago and deposited on the Gangplank. Then, relatively recently, it was eroded, transported, and deposited as gravel in Ash Creek.

Other recent sediments also present a puzzle. Nestled back in the hills and plastered against the side of the Ash Hollow slopes are some fine-grained and thinly layered lake deposits. Carbon-14 dating of organic matter in the lakebeds places their age at around 10,500 years old. How did lake sediments get deposited so high in the landscape? What caused the valley to fill with water at that time? Careful fieldwork has revealed shoreline deposits at about this elevation at other points around Lake McConaughy. The clear implication is that a lake or a cluster of lakes a fair bit larger than the present Lake McConaughy, perhaps 40 miles long and 100 feet deep, existed for centuries in this area. This ancient lake has been named Lake Diffendal after a geologist at the Nebraska Conservation and Survey Division who continues to contribute to the understanding of the geology of the state.

Lake Diffendal formed in a unique way. Common processes that form lakes include damming by glaciers, damming by landslides, the formation of tectonic depressions, or the abandonment of river channels. None of these processes makes sense here. Instead, the Sand Hills just to the north of the end of Lake McConaughy may have marched right into the valley and dammed the Platte River. Today, vegetation stabilizes the dunes, preventing them from migrating farther south. In Quaternary time, the sand dunes were particularly active. Sand migration must have been fast enough to overwhelm the river's capacity to carry it away. The valley would have

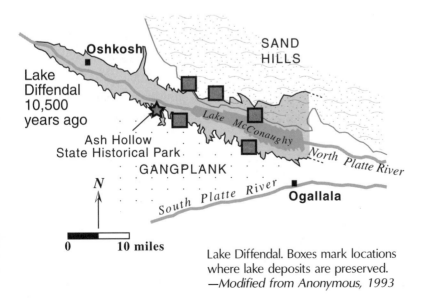

Lake Diffendal. Boxes mark locations where lake deposits are preserved.
—Modified from Anonymous, 1993

filled up with sand, forming a very wide and leaky dam. During subsequent wetter times, vegetation on the dunes slowed or stopped the movement of the sand, and flow rates were high enough to erode and remove the dam. Since the discovery of Lake Diffendal, geologists have identified other lakes dammed by dunes in Nebraska. Lake deposits in the Niobrara River valley can be found southwest of Valentine. Artifacts associated with the lakebeds indicate that Native Americans lived along the shores of Lake Diffendal. A bison kill site 6,000 to 9,000 years old was also excavated in terrace deposits of the Ash Hollow stream.

The Ash Hollow bluffs are carved into the Miocene-age Ash Hollow formation, which is the youngest formation in the Ogallala group. Capping some of the bluffs and forming very distinctive ledges are resistant horizons that are typically several feet thick. These ledges characterize the Ogallala group in this area. They are cemented together with calcium carbonate and contain a vast array of distinctive structures—well-preserved roots, insect burrows and galleries, and even hackberry seeds. Large, bulbous tubers that can be inches across are preserved yucca roots. These ledges are old soil profiles; from the top of the ledge down, the density of structures and calcium carbonate cementation decreases, until the sand and silt are stratified, undisturbed by the profusion of organic activity just above it. Preserved hackberry seeds and yucca roots let us infer that the environment wasn't very different from today's, although perhaps more seasonally arid. Calcium carbonate cementation and replacement characterizes soils of arid environments. Some geologists suggest a very shallow groundwater table at that time contributed to the calcium carbonate development. A well-developed ledge with abundant soil features is easily reached just west of the parking lot of the visitor center for the rock shelter. Watch out for the sharp spines of modern yuccas.

The multiple ledges of the Ash Hollow formation paint a picture of a cycle of deposition in a fluvial channel or floodplain environment followed by a prolonged period of stability—probably thousands of years—during which time soil formed in the surface material. A change in the course of the river or some other event caused that soil to be buried and preserved under new deposits. Given the number of ledges, this history must have been repeated often. While general climatic conditions and vegetation may have been similar to those of the present day, the fauna was not. Fossils of camels, rhinoceroses, and archaic elephants occur in the Ash Hollow formation.

On the footpath that winds down along the bluff exposures, you can see cross sections of well-developed channels within the Ash Hollow formation. These channels cut down several feet into slightly older strata and are filled with coarse conglomeratic material, some of which has well-developed crossbeds. The channel fill consists of poorly sorted sandstone and pieces of carbonate-cemented sediment similar to the heavily cemented hardpan layers of the ledge-forming soils. It appears to be locally derived material eroded from nearby deposits and redeposited in dry arroyos by flash floods during Ash Hollow time. Similar arroyos exist in western Nebraska today. Channels with coarse fill are not common throughout the Ash Hollow formation; this particular area may have been within a river system.

Four or more volcanic ash lenses that originated from volcanoes far to the west are interbedded with the Ash Hollow strata in this area. Two have been dated at 8 and 6.8 million years by fission-track methods, which measure scars left by radioactive decay. The two samples are separated vertically by 120 feet, and suggest a long-term sediment accumulation rate within the Ash Hollow formation of 100 feet per million years. This equates to 0.0012 inches per year. But clearly, much of the sediment accumulated at higher rates; a channel several feet thick may fill in just a matter of years or less. Loss of accumulated sediment to erosion accounts for some of the discrepancy, as do the periods of stability when soils developed. These hiatuses could easily have lasted for tens of thousands of years.

Where Ash Creek curves against the foot of the bluffs that contain the rock shelter, you can see the stratigraphic unit that lies beneath the Ash Hollow formation. It is composed of loess—homogenous light tan silt relatively rich in volcanic ash with some diffuse banding several feet thick. Clay development and a pattern of small-scale fracturing known as blocky structure indicate these diffuse bands are soil horizons, but of a different type than those in the Ash Hollow formation. Near the top of the unit, some calcium carbonate concretions occur. These loess deposits are assigned to the Whitney member of the Brule formation within the White River group and are some 30 million years old.

The Arikaree group that elsewhere occurs between the White River and Ogallala groups is missing here, a time gap—a disconformity—of close to 20 million years. The reason for the gap is not certain. Possibly the river channel system responsible for the overlying Ash Hollow formation scoured away the Arikaree sediments during the early, erosional phase of a cut-and-fill cycle. Or perhaps the Arikaree never accumulated to any great thickness here. The modern landscape may be part of a similar situation. The Platte River is stripping away the older Cenozoic record of sediments, but with time, it could deposit a whole new sequence of sediments during another fill cycle in the geologic future.

The Ash Hollow formation at Ash Hollow State Historical Park east of the rock shelter site. Arrow and dashed lines highlight channels. Resistant ledges and cap rock are old soil horizons.

U.S. 26 descends through Ash Hollow, then crosses the North Platte River floodplain. The river here is a broad, sandy, multi-channel, braided river system with abundant riparian vegetation. West of Lewellen, the road crosses Blue Creek, a groundwater-fed stream that originates just south of Crescent Lake in the Sand Hills. Crescent Lake is one of several hundred alkali lakes that are concentrated in a distinctive band across the Sand Hills. Groundwater flow connects these lakes, which formed when a sand dune dammed an old drainage system. Although dune forms are not obvious, eolian sand deposits occur along this stretch of the road.

West of Oshkosh, U.S. 26 runs along the foot of small bluffs of Ogallala group strata with some Brule formation strata at the base. A well-defined series of old channels, filled with sediment up to 300 feet thick, parallel the road at distances from several hundreds of yards to a mile. These represent an ancient river system similar to the Platte, parallel to but higher and offset to the north from the modern river. The ancient river can be traced from near Scotts Bluff to near the western end of Lake McConaughy. These channels are visible from the road but difficult to discern. The sediments in these channels belong to the Broadwater formation of Pliocene time.

Broadwater channel cut into the Ogallala strata in an area to the north of U.S. 26 some 8 miles east of the town of Broadwater.

Outcrops of Ogallala group along U.S. 26 with typical resistant ledges.

West of the town of Broadwater, the bluffs are farther north from the road, and Arikaree group strata begin to intervene between the Brule formation and overlying Ogallala group strata. From here west along U.S. 26 the Arikaree group gets consistently thicker until at Wildcat Ridge and Scottsbluff it reaches some of its greatest thicknesses. To the south of Wildcat Ridge, the Arikaree strata are again absent or much thinner. The basal Gering formation consists of fluvial sandstones. A major river channel complex in Oligocene time may have flowed through the area with the valley following approximately the length of what later became Wildcat Ridge, which may owe part of its origin to the greater resistance to erosion of the fluvial sandstones.

As you cross to the south side of the river at Bridgeport, you can see Wildcat Ridge to the south. The first lonely outposts of this feature are Courthouse Rock and its companion Jail Rock, small buttes near the confluence of Pumpkin Creek with the North Platte. A short drive south of U.S. 26, Courthouse Rock is a pleasant diversion and is easily accessible. The lower parts are the massive clay siltstones of the Brule formation. Above are Arikaree group sediments with ash horizons and sands with a plethora of crossbeds and channels. Crossbeds form whenever a flowing medium, such as water or wind, deposits sediments in ripples, sandbars, or dunes.

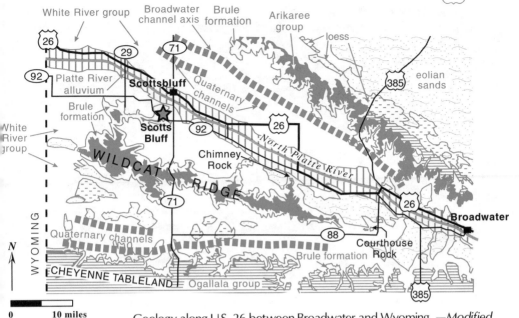

Geology along U.S. 26 between Broadwater and Wyoming. —*Modified from Burchett and Pabian, 1991; Swinehart and Diffendal, 1997*

Look for the fine laminae of the inclined sandy layers within the generally horizontal layers of the sediment.

Near Chimney Rock at South Bayard, U.S. 26 turns north, crosses the river and climbs the gentle valley wall before turning to head straight west to Scottsbluff. You may prefer to follow Nebraska 92, which follows a diagonal route along the North Platte to the same destination. This road stays on the south side of the river, passes Chimney Rock, and stays closer to Wildcat Ridge.

Chimney Rock is a pinnacle cut from Arikaree group strata similar to those exposed in the midsection of Scotts Bluff. While it looks quite delicate, it has persisted since pioneer times, when it was an important landmark along the Oregon Trail. A lightning strike was once observed to dislodge several large chunks of rock, adding lightning to the long list of agents of erosion. You can't walk up to the chimney to examine the strata, but you can see many of the same features at Scotts Bluff, about 20 miles to the west.

The town of Scottsbluff is dominated by its namesake just across the river to the south. The last landmark along the Oregon Trail in Nebraska, Scotts Bluff was a frequent stopping point for parties making their way west. A visitor center at Scotts Bluff National

View of Courthouse Rock and Jail Rock from the north. The Brule-Arikaree contact is the slope break in the lower third of the rock, with the bedding thinner and better defined in the overlying Arikaree sediments.

Close up of the Brule-Arikaree contact in Jail Rock with camera lens cap on the basal gritty layer in the Gering formation of the Arikaree group. Note the scour mark and crossbeds immediately above the lens cap. Massive clay silts of the Brule formation cross the bottom of the photo.

Monument presents information on the human and natural history of this area. A road climbs to the top of the bluff and walking trails bring you closer to the geology. Like the rest of the Wildcat Ridge area, the Brule formation of the White River group is exposed at the base of the bluff, especially in the small badlands area on the north side. The upper part provides spectacular exposures of the Gering and Monroe Creek–Harrison formations of the Arikaree group, and the trail allows you to walk up or down through this section. Scotts Bluff is one of the best places to see the geology of the Arikaree group up close. We present Scotts Bluff as a geolocality in the road guide for Nebraska 71.

It is apparent from a careful look at the topography or at topographic maps that the bluffs in Wildcat Ridge are aligned in two primary directions. The major direction is southeast/northwest, and the secondary direction is close to 90 degrees to the major direction. The alignment follows a regular array of mainly vertical fractures in the rock—a joint set. You can see these joints at Courthouse Rock, Chimney Rock, and Scotts Bluff. Tectonic stresses that affected the lithospheric plate of which these sediments are the uppermost layer formed the cracks after the sediments were deposited. As planes of weakness, joints control the directions of mass wasting and fluvial erosion, and thus help control the development of the drainage basin and landforms. These sets of joints are present in rocks across Nebraska and in neighboring states.

Whitish, massive ash layer in the well-stratified Gering formation of Jail Rock.

Continuing west from Scottsbluff, U.S. 26 follows the Oregon Trail along the floodplain of the North Platte River and on into Wyoming. At Mitchell, Nebraska 29 heads north through geologically interesting country to Agate Springs and Harrison, but U.S. 26 follows the relatively featureless lowlands to Wyoming.

The Cenozoic strata exposed along U.S. 26 tell a story of major river channels crossing and recrossing this general area over the last 30 million years. Early Cenozoic rivers carved channels into the underlying Pierre shale. Then, channels carved in Oligocene time began filling in with the Gering formation, followed by the rest of the Arikaree group, and those channels now stand above the landscape as Wildcat Ridge. Another episode of erosion removed the thinner parts of the Arikaree before rivers began to bury that landscape with Ogallala sediments in a series of cut-and-fill events that culminated in a broad surface, a small remnant of which forms the Gangplank. Then the Broadwater channels cut into the surface in Pliocene time, and finally the modern North Platte River system continues this erosional process. All of these river systems generally flowed from west to east. The Rocky Mountain uplifts to the west and north—Laramie Mountains, Hartville Uplift, and Black Hills—have funneled drainage through this part of western Nebraska since Paleocene time.

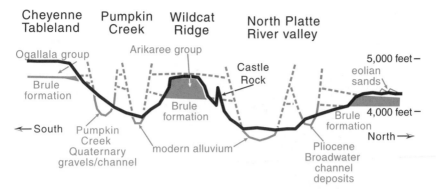

This cross section across west end of U.S. 26 shows the multiple channel deposits starting with the Arikaree sediments.

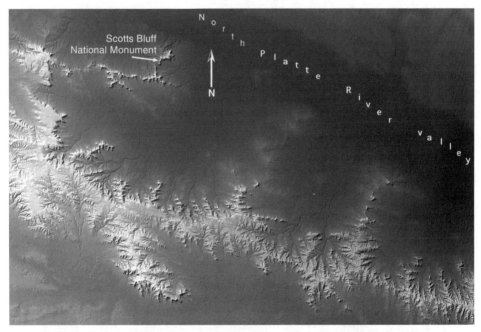

Shaded-relief map of Wildcat Ridge, a distinctive, linear feature that runs approximately parallel to the North Platte River. Image based on a digital elevation model of fifteen U.S. Geological Survey 7.5-minute quadrangles (Roubadeaux Pass in upper left to South Bayard in lower right).

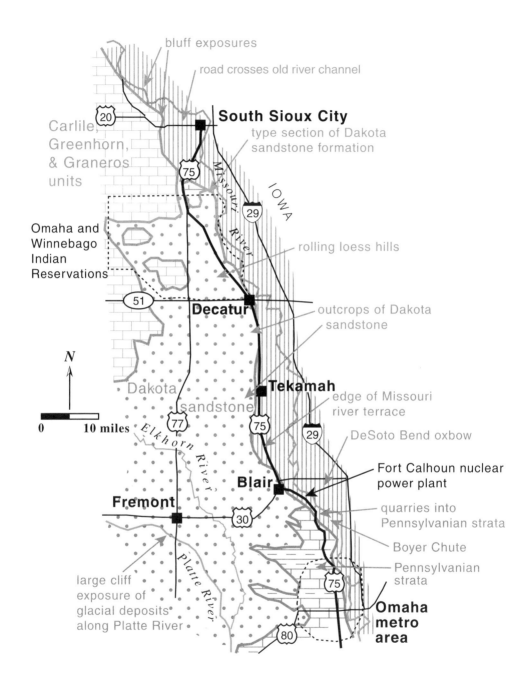

bluff exposures

road crosses old river channel

US 20

Carlile,
Greenhorn,
& Graneros
units

South Sioux City

type section of Dakota
sandstone formation

75

IOWA

29

Omaha and
Winnebago
Indian
Reservations

Missouri River

rolling loess hills

51

Decatur

outcrops of Dakota
sandstone

N

Dakota
sandstone

Tekamah

edge of Missouri
river terrace

77

Elkhorn River

75

29

0   10 miles

DeSoto Bend oxbow

Fort Calhoun nuclear
power plant

Blair

75

Fremont

30

quarries into
Pennsylvanian strata

Boyer Chute

Pennsylvanian
strata

large cliff
exposure of
glacial deposits
along Platte River

Platte River

75

**Omaha
metro
area**

80

Geology along U.S. 75 between Omaha and South Sioux City.
—*Modified from Burchett and others, 1975; Burchett and others, 1988*

U.S. 75
Omaha—South Sioux City
95 miles

U.S. 75 between Omaha and South Sioux City roughly parallels the Missouri River. In some places, such as near South Sioux City, it wanders out a good bit onto the floodplain and in other places, such as the southern half of the Omaha and Winnebago Indian Reservations, it climbs up into adjacent hills and bluffs overlooking the floodplain. For lengthy stretches, U.S. 75 runs right along the margin of or on old river terraces. This route provides plenty of opportunities to learn something about river landforms, processes, and history. In addition, the erosive action of the river created

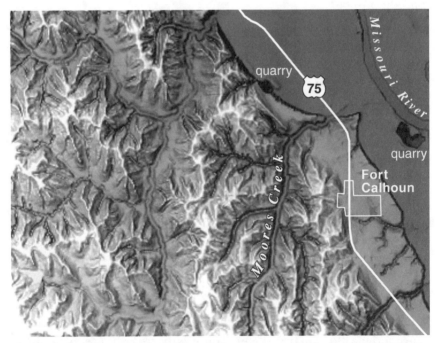

This shaded-relief map shows the Fort Calhoun terrace, the flat bench the town sits on. The northernmost quarry, visible from the road, is on a curved edge where the Missouri River recently carved a meander into the terrace deposit. Moores Creek, well demarcated in this image, has become entrenched since then because stream gradient increased when the terrace was cut into. A Missouri River meander bend associated with the time the terrace was an active floodplain cut the terrace's western margin (south of the town). Also note the level (white) areas capping some of the dissected loess hills—these are the remains of an older depositional surface on top of the loess that is now being dissected. The image, which has two-fold exaggeration of vertical relief, is a 30-meter digital elevation model of the Fort Calhoun 7.5-minute U.S. Geological Survey quadrangle.

some of the best available outcrops of the underlying sedimentary bedrock in eastern Nebraska, which is largely covered by Quaternary deposits. From south to north along this route successively younger bedrock units are exposed, from the Pennsylvanian carbonates and shales to the Dakota sandstones and Graneros shales of Cretaceous age.

On the outskirts of Omaha, U.S. 75 passes through the rolling loess hills that characterize much of eastern and central Omaha. Construction sites expose the light tan loess, but for the most part thick vegetation hides the underlying geology. The loess mantles glacial tills, which in turn overlie Pennsylvanian carbonates and shales—visible in some local abandoned and active quarries. The intervening Cretaceous strata are missing in the northeastern Omaha area.

About 5 miles north of the intersection with I-480, U.S. 75 descends from the loess hills to a Missouri River terrace, a flat area elevated above the modern floodplain of the Missouri. This surface was the river floodplain sometime in the past before another cut-and-fill cycle established the present landscape. The edge of old floodplain is at the toe of the hills to the west. Very low-angle alluvial fans spread out onto the old floodplain where streams exit the hills. The present drainage has downcut into the alluvial deposits and terrace, so the fans are no longer active and are covered with vegetation. The old floodplain is called the Fort Calhoun terrace, named after the small town directly on it. In town you will see signs directing you to Fort Atkinson State Historical Park where a reconstructed 1800s fort sits right on the edge of the Fort Calhoun terrace. The drop-off to the modern floodplain is quite noticeable.

Boyer Chute National Wildlife Refuge, established in 1996, is 3 miles east of Fort Calhoun on a signed road that drops down onto the modern floodplain. Boyer Chute is part the Back to the River program conducted by the Army Corps of Engineers and Nebraska. The program sponsors realized that for floodwater storage, wildlife and fish habitat, and recreational purposes, it would be useful to reconnect some subsidiary channels and wetlands with the main Missouri channel. The Army Corps had purposefully blocked these channels earlier for navigation and river control purposes, and sediment had partially filled the channels. At present the

Missouri River is a highly engineered river, with slanted stone groins along its length that keep the swifter water in the middle of the channel and prevent cutbank and point bar meandering dynamics. However, during floods, water has a nasty habit of choosing its own channel, sometimes favoring a minor channel over the main one. Because Boyer Chute is now reconnected to the river, it can store excess floodwaters. Entrance and exit control structures, largely hidden at the base of the channel, stabilize the flow through Boyer Chute.

Terraces along the Missouri River are not very common, and the stretch of U.S. 75 up to Tekamah has some of the best preserved examples. They provide geologists with important information about the river's history. These terraces are possibly related to Ice Age advances and retreats, which influenced both sea level and midcontinent fluvial dynamics, and may have formed during a drier and warmer interglacial period. Much geologic work on the history of the Missouri River remains to be done.

Just north of Fort Calhoun, U.S. 75 drops onto the modern floodplain. To the west you can see a quarry that cuts through the 20 feet or so of alluvium in this area and into the underlying Pennsylvanian strata. At the time of this writing one could see into the quarry from the dirt road to the north. Elsewhere the Missouri River alluvium is much thicker and deeper. This indicates that the depth of river erosion into bedrock was not uniform; deep valleys cut into the bedrock are now filled with river sediment.

Just south of Blair you can see two large building complexes. The first is the Fort Calhoun nuclear power plant. Reactor operation was initiated in 1973, and the plant is licensed until 2013 and is seeking a renewal until 2033. At present, as is the case for most nuclear power plants, the spent fuel is stored on-site. If Yucca Mountain or another site is given permission to receive high-level radioactive waste, then the high-level wastes may be stored permanently there. The plant was built near a river because water is used to cool the turbine steam, which is then recycled back into the reactor area. Barges were also a convenient way to transport some of the larger power plant components into place during construction.

The second building complex, closer to Blair, is an ethanol plant. About 10 percent of the nuclear power plant's output goes to the ethanol plant. Blair sits on a terrace and extends up into the loess hills.

We would be derelict if we didn't mention DeSoto Bend, a national wildlife refuge just across the Missouri River, seemingly in Iowa. The interior area surrounded by the curved lake is actually in Nebraska. In 1960 the Army Corps of Engineers produced the cutoff and shortened the river. Because the Corps—not the river—created the lake, this is an artificial oxbow. Around Thanksgiving time hundreds of thousands of snow geese and many bald eagles typically grace the refuge. The geese come in at dusk from foraging in surrounding fields and roost on the lake for safety. Perhaps the reason the geese migrate down the Missouri River corridor is to take advantage of the scattered oxbow lakes that provide safe haven from predators.

A hundred feet from the DeSoto Bend oxbow, the steamboat *Bertrand* was excavated in 1968. You can visit the pit from which it was taken. The cargo, which was headed up to supply the Black Hills gold rush, is well displayed at the refuge visitor center. The *Bertrand* sank in 1865 when it hit a log jam in the river channel. The Missouri, a very dynamic river, had significant navigational hazards until the Army Corps of engineers "tamed" it.

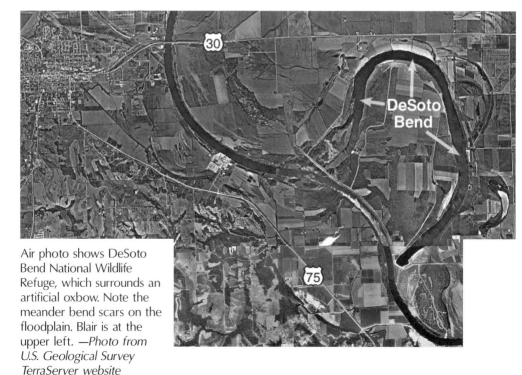

Air photo shows DeSoto Bend National Wildlife Refuge, which surrounds an artificial oxbow. Note the meander bend scars on the floodplain. Blair is at the upper left. —*Photo from U.S. Geological Survey TerraServer website*

Some 10 miles north of Blair at milepost 121, U.S. 75 follows the edge of another terrace that is lower, less modified by erosion, and possibly younger than the Fort Calhoun terrace. Just south of Herman, you can see two terraces. As you travel U.S. 75, you get a feel for the long, complicated history of the Missouri River. In the short term, channels meander and switch, producing oxbow lakes. In the long term, the river goes through cycles of cutting down a new valley and then filling it up with sediment and establishing a new floodplain. The terraces are all that is left of the former floodplains.

In the Tekamah area, you can find outcrops of Cretaceous Dakota sandstone, although none are easily seen from U.S. 75. Look for outcrops of the Dakota sandstone in the cliffs and hills to the west of the road between Tekamah and the southern border of the Indian reservation. Nebraska's one dinosaur bone locality is in the sandstone of this area. Some 12 miles north of Tekamah some small cliff exposures of the sandstone are visible from the road. Along this stretch, some large boulders in front yards are quartzites, granites, and other erratics from up north brought down by glaciers. In places, glacial deposits are preserved on top of the Dakota sandstone. For unknown reasons, loess deposits are sporadic to absent along this stretch.

North of Decatur, U.S. 75 climbs off the floodplain and terraces and into loess hills. The topographic relief is more marked—typical for areas of thicker loess deposits. Since the river is now meandering along the western side of the floodplain the highway must climb onto the bluffs. We recommend you stop at the Blackbird overlook where you can see the Missouri River with control groins along its banks, an oxbow lake, and old channel scars. An exhibit explains the history of the area, including Native American earthen lodges and the Lewis and Clark expedition. It is common to find remains of earthen lodges on the Missouri River bluffs, advantageous sites because of their vantage points.

Note that the loess hilltops are all at the same level to the west. This level represents some sort of old topographic surface. Streams dissecting that old surface created the hills. Former topographic surfaces are important features of Nebraska geology. This surface likely represents the top of a grassy plain that sloped gently away from the river and gradually built up as the grass captured windblown silt.

North of the Winnebago Indian Reservation, U.S. 75 follows Omaha Creek and its floodplain as it merges with the Missouri River floodplain. Exposures of the Dakota sandstones in the bluffs to the southeast are a type section locality—where the Dakota formation was formally described and defined in the literature. U.S. 75 continues towards South Sioux City diagonally across the floodplain, passing Crystal Lake, an old oxbow.

U.S. 75
Omaha—Kansas
91 miles

Traveling south on U.S. 75 through Omaha provides the opportunity to look at the landscape as urban and environmental geologists do. These scientists investigate how society can live more wisely on and with the landscape. Omaha stockyards—huge cattle holding and feedlot pens—are swiftly disappearing. Seepage of concentrated animal wastes can produce nitrate contaminant plumes in groundwater systems. As far as we know, this problem does not exist in Omaha, but as land use changes, it is important to consider any potential legacies hidden in the ground.

At the Chandler road exit, signs guide visitors to Fontenelle Forest, a large, private, nonprofit nature preserve situated on heavily dissected loess bluffs and Missouri River floodplain, complete with oxbow lakes.

Farther south, U.S. 75 descends onto the Papillion Creek floodplain. Here, Omaha has implemented an engineering and planning solution to flooding. Urban runoff has increased the flooding potential of Papillion Creek, which passes through much of Omaha. Multiple government agencies spent millions of dollars to channelize the stream and build levees along its margins. Development is not permitted on the floodplain, which is used instead for recreational green space. A trail system follows the stream from the interior of the city all the way downstream to the Missouri. From the Papillion Creek floodplain, you can see Offutt Air Force Base to the east, and because it has dealt with large quantities of potential contaminants, there are contaminant clean-up operations here.

Continuing south, U.S. 75 crosses a small divide between Papillion Creek and the Platte River and then descends onto the Platte River floodplain. This stretch of road crosses the edge of the buried Humboldt fault zone, which stretches from this point south into Kansas. It follows a major and much broader geophysical anomaly known as the midcontinent gravity high. The pull of gravity is a little bit greater here due to the basalts of a very old buried rift structure some 2,500 feet below the surface, the Midcontinent Rift. Experience elsewhere teaches us that such old rifts in the interior of continents are prone to earthquakes. The New Madrid zone in Missouri, which triggered three large earthquakes in 1811 and 1812 (magnitudes of approximately 7.5, 7.3, and 7.8), is the best example of this. The Humboldt fault zone in Nebraska is seismically active and poses some risk, although geologists believe the risk is substantially less than that associated with the New Madrid area. The most notable earthquake along the Humboldt fault was in 1935 near Tecumseh and caused minor damage—cracked plaster, broken windows, and a few toppled chimneys.

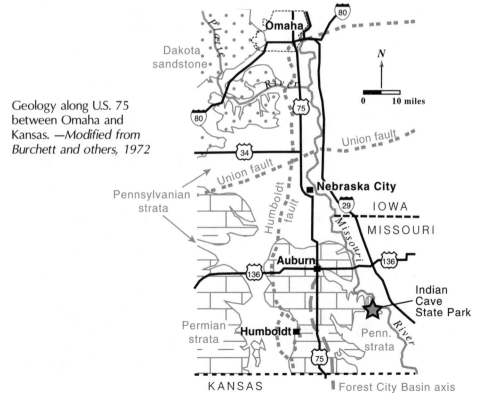

Geology along U.S. 75 between Omaha and Kansas. —*Modified from Burchett and others, 1972*

Where the Midcontinent Rift is exposed in Michigan, there are organic-rich black shales associated with it. These deposits can be good source rocks for oil and gas. The subsurface Midcontinent Rift sediments were drilled in Iowa, but no oil was found.

Just upstream of the bridge across the Platte River (but not visible from U.S. 75), a well field along the sandy banks of the Platte draws water for the southern Omaha metropolitan area. The water actually comes from the river, but pumping pulls the water through the sands, which filter it. Groundwater is a crucial resource to Nebraska, and shallow alluvial aquifers, such as the one beneath the Platte River floodplain, are common municipal water sources.

A number of small lakes adjacent to the Platte River fill old gravel pits, and their surface basically represents the surface of the shallow water table. Gravel and sand are important construction commodities for an urban area. Much of the expense in mining gravel and sand is transportation, so proximity to urban areas is an important economic factor in the industry. Following excavation of gravel pits, the lakes become attractive sites for cottages and houses. Contaminants can more directly enter the alluvial aquifer through the lakes. From a historical perspective it is interesting to think about what these buried gravel deposits represent. At present the Platte moves mainly sand, and the substantial buried gravels suggest a more vigorous river in the past.

As U.S. 75 climbs out of the Platte River valley it passes through vegetated "rolling" hills typical of the very eastern portion of Nebraska. The hilltops are not rounded but rather share a common, relatively flat surface. A myriad of streams dissected this older surface into hills. Unlike the surface in the Indian reservation on the northern portion of U.S. 75 with its underpinnings of loess, here glacial till largely makes up the upper portions of the hills. This may be an old glacial surface, perhaps left behind from the retreat of the continental ice sheet more than 600,000 years ago.

Where U.S. 75 crosses Weeping Water River, you can see a green structure with a satellite dish on its top. This is a U.S. Geological Survey flood gauge, which monitors the amount of flow or discharge in the river. A large system of such gauges around the country collects important data. Data of past flow histories helps in the prediction of the likelihood of future flooding. By analyzing the data, hydrologists estimate the size of a one-hundred-year flood for a given river stretch. Such information is used in city zoning and planning, setting of insurance premiums, and construction design of bridges, dams, and flood and erosion control structures. Stream discharge information is now available free of charge from the U.S. Geological Survey website.

Nebraska City boasts a number of tourist attractions, foremost of which is Arbor Lodge State Historical Park, where visitors can walk through the Morton Manor. J. Sterling Morton, of the Morton Salt family fame, was U.S. Secretary of Agriculture under President Grover Cleveland. Morton established nurseries and orchards so the state of Nebraska began Arbor Day in his honor.

At the west footing of the bridge crossing the Missouri River into Nebraska City, a good section of rock is preserved. These Pennsylvanian Waubansee group limestones, shales, and sandstones also form cliff outcrops tens of feet high along the railroad tracks that parallel the river. Brachiopods, fusulinids, and crinoids in the limestones give testament to the sea that used to shift across the midcontinent back in Pennsylvanian time. Some interbedded sands are particularly rich in detrital muscovite, a

Outcrops just to the north of the bridge abutment where Nebraska 2 crosses the Missouri River east of Nebraska City. Ledges are limestones interbedded with siltstones and sandstones of the Pennsylvanian Waubansee group.

light-colored mica, which is a weathering product of granitic or metamorphic rocks. It is probably derived from some bedrock exposures in mountains. In this part of the United States, we usually look west towards the Rockies for sediment sources, but the Rockies did not exist in Pennsylvanian time. At that time, the Appalachian Mountains were forming and were several times higher than they are now. Huge deltaic plains used to extend from the Appalachians all the way into eastern Nebraska. The Pennsylvanian strata are much coarser and thicker closer to the Appalachians to the east. Overlying the Pennsylvanian rocks are benches of Quaternary loess.

U.S. 75 descends onto the Little Nemaha River floodplain just north of Auburn. Many of the rivers in this region have a clear linear northwest-southeast trend. A regular array of fractures in the underlying bedrock may control the drainage pattern. Water seeps into fractures, quickening the weathering and erosion along these paths, which grew into stream valleys.

From Auburn south, surficial deposits are thinner and roadcuts and other exposures of bedrock are more common. U.S. 75 follows the axis of the Forest City Basin, where a great thickness of sediment accumulated in Permian time. The basin is related to the long-lived tectonic activity along the Midcontinent Rift. As is typical in a basin, somewhat younger rocks are exposed in its interior—in this case Permian strata of the Admire and Council Grove groups. Ancient soils preserved in these sections indicate the surface was periodically above sea level, while limestones with marine fossils indicate inundation by the sea. Some of the carbonates have structures suggestive of evaporite formation. Mica-rich sands are much more common in Permian strata than in the underlying Pennsylvanian strata. The picture that emerges is of a broad, arid, and flat continental interior that inland seas swept over and retreated from as the Appalachian Mountains shed sediment westward. The deposits left from these cycles of advance and retreat are known as cyclothems.

The Forest City Basin has some notable oil fields, mostly in Kansas. Near where U.S. 75 crosses the South Fork of the Big Nemaha River is the 680-acre Dawson oil field. Maximum annual yield was some 117,841 barrels in 1955. The reservoir is in Devonian and Ordovician strata arched in a subsurface anticline, which is likely related to Pennsylvanian reactivation of structures in the buried Midcontinent Rift.

On the higher hilltops in the area, thin tills are remnants of the now-dissected Pleistocene surface. Some of the clasts in the till are the familiar red Sioux quartzites of South Dakota and Minnesota. The glacial deposits stretch all the way south into Kansas.

Where U.S. 75 crosses it, the South Fork of the Big Nemaha River is confined to a linear channel created by the Army Corps of Engineers, but remnants of the old meandering channel exist to the north. Perched 10 to 12 meters high on the south valley wall are some even older thin alluvial deposits sitting on an old river bed erosional surface known as a strath terrace.

Two geologists, Rolfe Mandel and Art Bettis, studied landscape evolution here. Roughly 10,000 years ago the Nemaha River was flowing across Paleozoic bedrock. Older loess-mantled terraces some 15 meters or higher formed the valley walls. From 10,200 years ago to about 7,000 years ago floodplain deposits built up. Between 7,000 and 4,800 years ago sediment built up slowly enough that a thick soil developed in the floodplain sediments. Around 4,800 years ago, the river incised the floodplain, cutting down to the bedrock in some places. This channel then filled in until some 3,600 years ago. Since then, incision has occurred three times, including the present period, with limited infilling of the channel areas and some deposition on adjacent floodplain in between. Such a detailed study displays the complex cut-and-fill cycles that characterize fluvial systems.

Indian Cave State Park

Set in the rolling and wooded hills adjacent to the Missouri River north of Falls City, Indian Cave State Park is named after rock overhangs in the river bluffs. These easily accessible bluffs are a major geologic attraction. The "caves" are not dissolved by solution but rather eroded in iron-stained sandstones, pebbly sandstones, and some siltstones, with good crossbeds and internal channel forms. Native American carvings are preserved in the cliffs.

These strata resemble the Cretaceous Dakota sandstone but are older—primarily Permian in age. The type of sediment and sedimentary structures, coupled with a lack of lateral continuity indicate these are a part of a large channel deposit. Some horizons show flat clasts of fine-grained, laminated clays embedded within the sands. These clasts are referred to as rip-ups because floods rip up pieces of dessicated, mud-cracked clay. The relatively fragile clay clasts get deposited and preserved quickly. A close inspection of the base of the cliffs reveals some small outcrops of the thinner-bedded siltstones, sandstones, and limestones that characterize the upper Pennsylvanian in southeast Nebraska. These are the strata the channel was cut into. Oscillating sea level produced cyclothems in Pennsylvanian time. However, superimposed on top of these cycles of shale and limestone layers is a longer-term regressive trend. This is evident as an increase in the proportion of terrestrial strata relative to marine strata higher in the Pennsylvanian section, as occurs here.

Profuse coal deposits of Pennsylvanian time helped fuel the Industrial Revolution in England. The conditions for coal formation were prevalent not only in England, but also in an area west of the Appalachian Mountains. These coals formed from organic matter deposited in swamps along the edge of the large wedge of sediments shed from the Appalachian Mountains. Such coals extend just into Nebraska. Several hundred feet south from Indian Cave was a coal shaft, where a Mr. Deaver mined the coal to heat his house. This coal was of relatively poor quality.

Some 100,000 bushels of coal were mined from Pennsylvanian strata in the late 1800s from the very southwestern corner of Richardson County in which Indian Cave resides. The coal seam was 18 to 30 inches thick, yielded little ash, and was used locally. Early reports on Nebraska often mentioned the promise of coal deposits as a selling point to entice settlers and industry. While coal has been encountered elsewhere in drill holes, it is not extensive or shallow enough to mine.

Indian Cave is an erosional rock shelter and was not formed by solution. The Missouri River probably scoured out the cave when the water flowed directly against the sandstone cliffs. In that case, the cave would have formed before the river cut down to its present

Cave eroded into Permian sandstones at Indian Cave State Park.

level, perhaps during or before the last Ice Age, making the cave older than 12,000 years. However, a companion erosional process may have contributed to the cave's formation and may be still modifying it. The channel sands the cave developed in are weakly cemented together. Such sandstones transmit groundwater flow, and water seeps out of the sandstone at the base of the bluffs. Chemical reactions and freeze-thaw action at the rock surface where the seep occurs can break the cement bonds and loosen sand grains. Runoff, wind, and occasional high floods can carry away those grains, gradually enlarging the cave back into the cliff in the direction the water is seeping through the rock. Loose sand on the floor of the cave may have eroded in this manner. Such a process is responsible for caves in sandstones elsewhere, including those at Mesa Verde National Park in Colorado.

Please resist any urge to carve your initials in this sandstone. Native American carvings, mainly of wildlife, are nearly obscured by all the recent illegal carvings. This vandalism has partly destroyed some petroglyphs. Not enough remain to characterize their style or to assign an age to them. In addition, natural weathering processes associated with the moist environment are degrading and destroying these carvings at a much faster rate than images chipped through the desert varnish of many petroglyph sites of the more arid Southwest.

U.S. 83
North Platte—Valentine
130 miles

U.S. 83 traverses the Sand Hills, so you might expect it to be geologically monotonous. However, if you pay attention to the characteristics of the dunes, you discover that different dune types occur in different areas. The sand sea has an internal structure organized by geomorphic processes. Driving across this sand sea for hours will also give you a true appreciation of its magnitude. Four distinctive rivers break the monotony, the South, Middle, and North Loup Rivers and the Dismal River. The Sand Hills give them their distinctive character. The river valleys also provide occasional glimpses at what is underneath all that sand.

The point at which U.S. 83 climbs north out of the Platte River valley is a fairly well-defined slope break. After passing some thin loess deposits, the highway traverses a patch of simple, domal dunes that are muted and relatively thin. The top, organic-rich soil horizon is better developed here than elsewhere, as are erosional features, suggesting these domal dunes are a more degraded part of the Sand Hills dune complex. You can also see a fair number of blowouts, evidence of modern wind erosion.

After passing through about 8 miles of dunes, the surrounding area becomes quite flat and is covered by cropland and trees. This area, known as the Garfield Table, is clearly not in the Sand Hills. It is underlain by Pleistocene loess and Pliocene sands and gravels. The highway drops gradually onto the very broad floodplain of the South Loup River as you travel north of Stapleton.

North of the South Loup River, the Sand Hills are well developed, and U.S. 83 cuts across long dune forms with crests oriented roughly east-west. These are the classic, large barchanoid-ridge

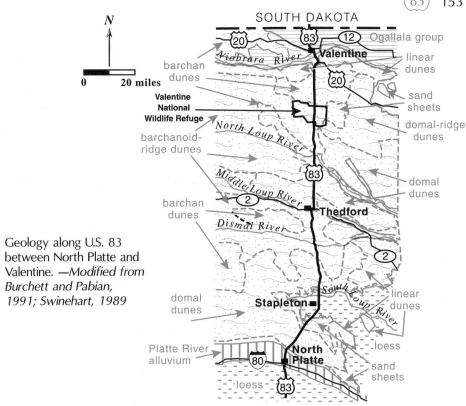

N

0   20 miles

SOUTH DAKOTA

20   83   12   Ogallala group

Niobrara River   Valentine   linear dunes

barchan dunes   20

Valentine National Wildlife Refuge   sand sheets

North Loup River   domal-ridge dunes

barchanoid-ridge dunes   83

barchan dunes   *Middle Loup River*   domal dunes

2

Dismal River   Thedford

Geology along U.S. 83 between North Platte and Valentine. —*Modified from Burchett and Pabian, 1991; Swinehart, 1989*   2

domal dunes   Stapleton   *South Loup River*   linear dunes

Platte River alluvium   80   North Platte   loess

loess   83   sand sheets

dunes, sometimes known as transverse dunes, that often come to mind when you think of dunes. Scouring wind shaped their gently sloping north face. Sand, sheltered from the wind in the lee of the dune, built up the steeper south face. It is very clear from this morphology that the prevailing wind direction during dune growth and migration was out of the north. In some blowouts along the road, you can see the crossbedded internal structure of the dunes.

Where U.S. 83 crosses it, the Dismal River valley is fairly narrow and well defined in comparison to the Loup River valleys. The river flows through a remote and undeveloped area, and its relatively fast currents, tight turns, and narrow channel are popular with paddlers. A turnout on the north side of the river provides an excellent overlook. Upstream for some 20 miles, the valley closely follows the area between two particularly well-developed, large barchanoid–ridge dunes. While the wind was transporting sand from north to south, the river transported sands eastward. Sands blowing in from the north would have to cross the river to form dunes to the south. When the river was active, it probably

View to the southwest of the Dismal River and U.S. 83 bridge. Note the dune forms on the horizon.

intercepted the windblown sand and carried it away to the east, starving the downwind area of sand. At the time the large dunes to the south formed, the river may not have existed yet. Or the sand dunes may have overwhelmed the river, filled the valley, and marched uninterrupted to the south. The fact that the valley follows the trough between dunes suggests that the dunes pre-dated the Dismal River and played a role in the formation of the valley. Flow of groundwater into the river drains the surrounding higher sands and lowers the adjacent water table. A cluster of lakes marks the headwaters of the Dismal River. Lakes may have existed here, too, before the incision by the Dismal River drained them by lowering the water table.

Scientists appreciate the opportunity to put numbers on geologic processes, and such an opportunity exists in calculating how rapidly the Dismal River incised its valley. Some radiocarbon ages from organic matter in the sediments beneath the dunes indicate that the valley may have formed in the last 3,000 years. The valley is incised almost 100 feet below the base of the sand deposits, so

The steep, southern face of part of large barchanoid-ridge dune north of the Dismal River.

the river cut down about 3.3 feet per 100 years during the 1,500-year period. While we wouldn't suggest waiting around in hopes of seeing the incising progress, this is a fast rate for a geologic process. But given the constant supply of groundwater and the erodability of the sediment, it is not surprising. The incision of the valley probably progressed in an upstream direction—the headward erosion of the river cut westward into the Sand Hills over time and still does.

The Dismal River is unique in several ways other than its young age. For its size in terms of discharge, it is relatively short (only some 80 miles long), has very few tributaries, and has a relatively constant flow. The constancy of its flow is apparent from the relatively steep banks with abundant bank vegetation. Evidence of scouring by seasonal or storm-related floods is absent. A U.S. Geological Survey monitoring station exists on the river near the U.S. 83 bridge, and average daily flows only vary from 130 to 200 cubic feet per second. The flow variations of most rivers differ by several hundred percent if not by an order of magnitude. The reason for this constancy and for many of the other unique traits is well known: the Dismal is fed primarily by groundwater, with the majority of the flow originating from the aquifers within a thousand feet on either side of the channel.

While some water seeps out of the adjacent Sand Hills, the existence of "boiling" sand springs along the river indicates a significant amount comes from a deeper source. These springs emerge from bowl-shaped depressions in the alluvium of the river valley, usually within 100 feet of the river. The depressions are up to 30 feet in diameter, and at their center have a vertical conduit where water is moving up at a rate fast enough to cause the sand to move around. Though the springs look like they are boiling, they are not hot. Flow is not constant—it pulses. Weights have been lowered through the turbulent sands to a depth of 130 feet. Such depth suggests that some of these conduits are connected to the underlying Ogallala aquifer. The well-defined sides of these

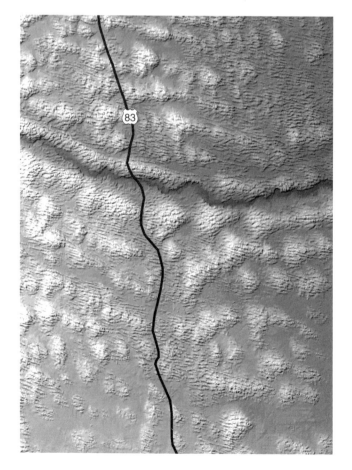

This shaded-relief map shows topography where U.S. 83 crosses the Dismal River. You can see smaller dune forms superimposed on larger composite dune forms, both with an approximate east-west axis. Note how the river meanders between two long, composite dune features and has no significant tributaries. Image is from a composite 30-meter digital elevation model of four U.S. Geological Survey 7.5-minute quadrangles (Halsey Southwest, Happy Hollow, Mudd Lake, Thedford Southeast).

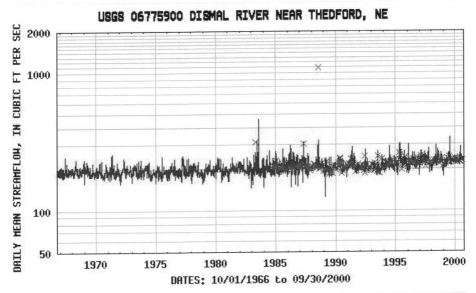

USGS 06775900 DISMAL RIVER NEAR THEDFORD, NE

DATES: 10/01/1966 to 09/30/2000

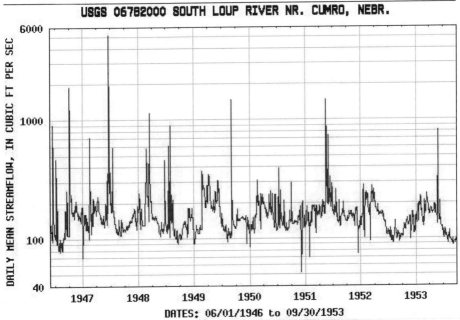

USGS 06782000 SOUTH LOUP RIVER NR. CUMRO, NEBR.

DATES: 06/01/1946 to 09/30/1953

These graphs show the seasonal history of water flow in the Dismal and South Loup Rivers downstream of where U.S. 83 crosses both rivers. The Dismal River, embedded within the Sand Hills and largely groundwater fed, shows much less variation in flow than this portion of the South Loup just south of the Sand Hills, which has a supporting drainage network and is influenced to a much greater degree by storm runoff.

conduits are firmly packed sand or, at greater depths, clayey silt. How these vertical, pipelike conduits formed in the alluvium is unclear. They challenge our normal perceptions of how groundwater flows through sand; we usually think of groundwater flow as evenly distributed through all of the sand, but here, groundwater flow is concentrated in distinct conduits.

The groundwater also does not simply flow downward from the surrounding hills. Instead, it flows to deeper levels before rising back up, following a path with significant vertical relief. The topography of the ground surface and the position of the water table drives the flow. The elevated water table within the adjacent hills forces the water down before it rises beneath the valley. The material the river flows over includes not only its own modern alluvium, but also Pliocene alluvium up to 60 feet thick from older and broader river systems.

A couple of miles north of the Dismal River, the dune forms become less regular and more complex. U.S. 83 drops into the broad Middle Loup River valley and merges briefly with Nebraska 2 at Thedford before continuing northward. In the area between the

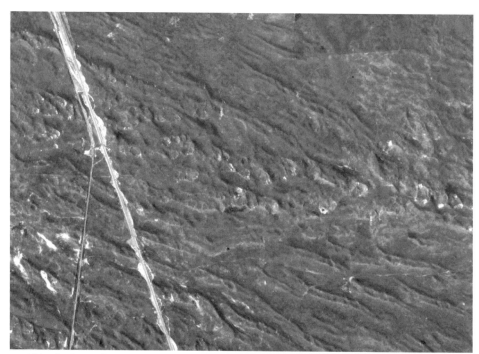

This air photo shows U.S. 83 crossing barchanoid-ridge dunes several miles north of the Dismal River. The wind direction when the dunes formed was from the northwest. Some of the lighter areas are sand exposed in blowouts. —*Photo from U.S. Geological Survey TerraServer website*

Middle Loup and North Loup Rivers, the sand dunes are not well defined and consist of simple domal dunes and sand sheets. The North Loup River may have captured the windblown sand and starved the area to its south of sand. North of the North Loup, the sand dunes are a domal-ridge type and are larger than south of the river.

The Valentine National Wildlife Refuge encompasses a distinct concentration of lakes and wetlands that formed in the interdune depressions. These groundwater-fed lakes reflect the emergence of a very shallow groundwater table. To the east, some of these wetlands feed Plum Creek; this concentration of lakes may be an area where sand dunes filled some of the valleys cut by an older Plum Creek drainage system. These lakes and wetlands are an important habitat for wildlife, especially birds. Some 260 species of birds have been observed here, and flocks of pelicans are not an uncommon sight.

Flat cropland marks the local northern edge of the Sand Hills and the point where U.S. 83 emerges onto a tableland of Ogallala sediments. The route joins U.S. 20 and crosses the deeply incised Niobrara River valley before continuing into Valentine.

U.S. 136
Missouri—Arapahoe
246 miles

Rich in both geologic and human history, U.S. 136 provides a cross section of the Great Plains. From the Missouri River, it quickly climbs up onto the Pleistocene glacial surface, which is now dissected into rolling hills. About where it crosses the Oregon Trail, near Fairbury, the road reaches the western limit of the glacial till, and rises and falls through four major units: the Dakota sandstone, Niobrara chalk, and Pierre shale of Cretaceous age and the Ogallala sediments of Tertiary age. Above these units are Pleistocene gravels and the ubiquitous loess deposits. But this complete sequence is almost never present in one place. While the marine deposits of late Cretaceous time are widespread and retain similar characteristics over long distances, the Tertiary and Quaternary units are nonmarine and quite discontinuous, removed to varying degrees by erosion and in some places entirely absent. The western half

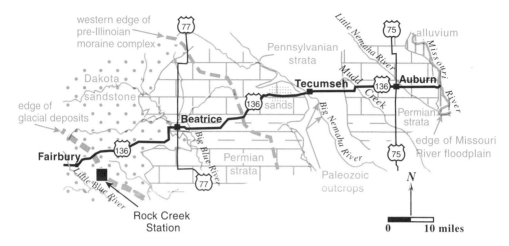

western edge of
pre-Illinoian
moraine complex

Pennsylvanian
strata

Little Nemaha River

alluvium

Missouri

Dakota
sandstone

Tecumseh

Auburn

edge of
glacial deposits

Beatrice

sands

Permian
strata

Mudd
Creek

Missouri River

Big Nemaha River

Fairbury

Big Blue River

Permian
strata

edge of Missouri
River floodplain

Little Blue River

Paleozoic
outcrops

N

Rock Creek
Station

0   10 miles

Geology along U.S. 136 between Missouri and Fairbury. Glacial till
blankets the surface. —*Modified from Burchett and others, 1972;
Dreeszen and others, 1973*

of U.S. 136 follows the Republican River valley, where surface
hydrology, fluvial geomorphology, and groundwater are geologically and politically important.

U.S. 136 enters Nebraska at Brownville, a small river town that
maintains some of the flavor of its early history to encourage
tourism. Nearby is the Cooper Nuclear Power Station, one of the
two nuclear power stations operated in Nebraska by the Nebraska
Public Power District. From Brownville to Tecumseh, outcrops are
relatively rare. The most evident geological features are the well-
developed floodplains of the Little and Big Nemaha Rivers and the
well-vegetated, rolling hills cut into the glacial till. In 1935 Tecumseh
was the site of one of Nebraska's largest historical earthquakes,
which was probably a low magnitude 5 on the Richter scale. The
Humboldt fault zone shifted, cracking plaster, breaking windows,
and toppling a few chimneys.

Outcrops of glacial till with quartzite, granite, and chert clasts
are exposed in cuts behind the small mini-mall on Nebraska 50
just south of the intersection with U.S. 136. The till, fairly representative of the deposits in this area, is a reddish brown, and surface
exposures often display an irregular array of cracks. When the clay
in the till gets wet, it swells, and when it dries, it shrinks. The
change in volume causes the cracks. The mixture of extremely

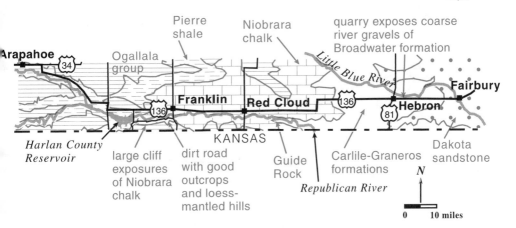

Geology along U.S. 136 between Fairbury and Arapahoe.
—*Modified from Dreeszen and others, 1973*

fine-grained clays with boulders identifies this material as till. Only glaciers carry sediment with such a wide range of grain sizes, transport it significant distances, and deposit it all together. The glacier transported the distinctive quartzite clasts from far to the north. Dark brown soils are developed within this till. Outcrops of the till are common from Tecumseh to Fairbury.

About 6 miles west of Tecumseh, the hilltops are composed of different material, a fine-grained white sand. The till lies beneath these sands, which must be Pleistocene or younger alluvial or lake deposits. The sand is high above the present drainage systems so modern streams could not have deposited it. Just east of Fairbury is the westernmost edge of the till.

Crossing the Little Blue River just west of Fairbury, you can see light and dark layers in Holocene alluvium in the exposures along the riverbank. Levees built along the river attest to concerns about flooding. Some 2.7 miles west of the river on the south side of U.S. 136, an exposure of coarse gravel is perched up on the side of the valley. Such gravels represent the deposits of some ancient vigorous river that flowed swiftly enough to move such coarse debris. Its position in the landscape indicates it is not associated with the present drainage and is likely Pleistocene in age. If you looked at the pebbles that make up the gravel, you'd find pieces of granite, a lot of quartz from veins in igneous or metamorphic rocks, some gneisses, and a few schists. Clasts of the Sioux quartzite from far to the north are missing from the gravel even though they exist in the nearby glacial till. Thus, the clasts probably originated from the Rocky Mountains to the west rather than from the local glacial till.

A few miles farther west, cuts at the tops of hills expose Peoria loess in the upper portions and a dark brown clay-rich unit below. Calcareous concretions are common in the loess, and the recent soil developed in the upper portion has some calcium carbonate development reflecting dry climatic conditions. The brown, clay-rich material below lacks coarse grains and is west of any glacial till. It may represent Pleistocene lake deposits.

About 8 miles west of Gilead, a large private gravel pit on the south side of U.S. 136 provides a rare window into the geology. In the quarry wall, tens of feet of light-colored sands with large-scale crossbeds are overlain by 10 or more feet of gravel, which in turn is overlain by stratified loess near the top of the hill. Such coarse deposits and large crossbeds characterize river channel deposits.

A roadcut on U.S. 136 less than a mile to the east exposes the upper gravel. The gravels contain pieces of volcanic rocks, whose likely origin is in Colorado, and also pieces of anorthosite, a very distinctive and uncommon igneous rock made of plagioclase crystals. A large anorthosite body exposed in the Laramie Mountains of Wyoming is the most likely source for this gravel. Other gravel deposits in south-central and western Nebraska also have anorthosite pebbles in them. While these may represent deposits from different channels and may not have been connected in time and space, they are collectively called the Pliocene Broadwater formation. The coarse gravels in roadcuts and in the quarry were probably deposited by an ancient version of the Platte River, which swung much farther south than the modern Platte River and had tributaries in both Wyoming and Colorado.

At the town of Guide Rock, U.S. 136 follows the northern edge of the Republican River valley. Guide Rock, not so much a rock as a loess bluff of modest size, overlooks the valley on the south side of the river. Historical markers at the bluff indicate that it gained its name because it was a holy site to the Pawnee, who inhabited a large village nearby in the late 1700s.

Red Cloud is known as the home of Willa Cather, and some nine thousand Cather fans per year make pilgrimages to the town. Six of her twelve novels were set in Red Cloud, and the house she

Gravel quarry wall along U.S. 136 west of Gilead. Note the erosional contact between lower white sands and gravels with large crossbeds and overlying gravel and sand beds.

lived in is a state historical site. In the Red Cloud area, sporadic loess deposits overlie gravels, which overlie the Cretaceous Niobrara formation. All of these have been dissected by the Republican River and its tributaries, which have deposited modern river alluvium in the valleys. Just east of Red Cloud, low outcrops of white chalk and shale along the side of U.S. 136 reveal the presence of the Niobrara formation, deposited in the Western Interior Seaway of Cretaceous time. In roadcuts just east of Riverton, a gravel rich in white chalk clasts mixed with granite clasts sits on top of the Niobrara chalk. The gravel may be the Broadwater formation, or some more recent Republican River alluvial terrace deposits. On the west side of Inavale are more extensive outcrops of the Niobrara formation that include light gray shales. You can find shells of inoceramid bivalves in the limy layers here. In some of the upper

Air photo of the Republican River valley. Inavale at middle right. —*Photo from U.S. Geological Survey*

parts of the Niobrara formation exposed in Kansas these fossil shells may be up to 3 feet across. Similar Niobrara chalk is also present in northeastern Nebraska.

Several miles east of Franklin, you can see the Franklin irrigation canal on the north side of U.S. 136. This canal cuts through rounded hills of thick loess deposits. Consider the effort it took to dig these canals straight through the loess hills and you can get some idea of the importance of irrigation for agriculture in this area. Loess thicknesses are well in excess of 40 feet here. You can see Pliocene Broadwater gravels low in some roadcuts along this stretch of the highway.

At Franklin, we suggest that you depart from the highway and take the gravel road that skirts the edge of the floodplain on the south side of the river and leads to the south end of the dam that impounds Harlan County Reservoir. Some of the better outcrops in this area are along this road. Where the road cuts through modern hills, you can see loess mantling ancient hills of Niobrara chalk and shale. In some places, where the valleys of tributary streams meet the floodplain, outcrops expose loess all the way down to the level of the modern floodplain. The tributary valleys have carved into an ancient loess-filled valley that was initially carved into the Niobrara sedimentary rocks. Where the road crosses Lost Creek,

Loess mantles an old hill of white Niobrara chalk with horizontal bedding along a gravel road on the south side of the Republican River valley south of Franklin.

there is a cliff of loess. But, as you round the corner, roadcuts expose a buried hill of Niobrara chalk.

Just east of the intersection of the gravel road with the road from Bloomington, a large cliff exposes 65 feet of the Niobrara formation. Note the colors of the Niobrara chalk and shale here. Weathering rinds and rims along the upper portion of the outcrop and along horizontal and vertical fractures show that weathering colors the chalk beige. Less weathered material is gray, and the freshest material is very dark gray. Organic material imparts the dark color. Weathering processes such as oxidation and bacterial action degrade the organics, leaving the rock lighter in color, due to the remaining carbonate material. Thus the Niobrara formation exposed in this cliff is quite rich in organics. It served as a carbon sink during Cretaceous time as billions of tiny organisms extracted carbon dioxide from the sea and the atmosphere to build calcium carbonate skeletons (the chalk) and their bodies (the organic material). Upon dying, they sank to the seafloor and the accumulating sediment locked the carbon in place. In the middle of the Western Interior Seaway, where clays, silts, and sands rarely reached, the majority of the sediments were the products of biological activity. This boom in the accumulation of carbonate and organic-rich sediments was part of a global phenomenon.

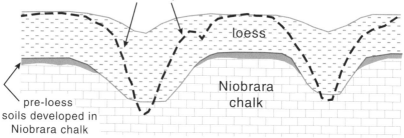

This schematic cross section shows a buried hill inside a modern hill. An older surface with valleys and associated soil horizons was cut into the Niobrara chalk. Loess deposits then draped over and buried this old landscape, muting the valleys. Erosion and dissection along these valleys cut down through the loess and into the chalks to form the modern topography.

You may notice a faint, pungent odor while standing at the outcrop or from a freshly broken specimen. What you smell is the hydrocarbons in the rock. The organic content is of interest to companies that explore for petroleum.

In the lower half of this outcrop there are a few distinctive thin, white layers of volcanic ash carried by the wind from large volcanic eruptions far to the west. The volcanic ash, which originally consisted of tiny fragments of pumice, a form of volcanic glass, has been altered to a waxy clay called bentonite. Because these ash layers represent a very brief event in geologic history and were deposited over a large area at virtually the same time, they are very useful as time markers. You can find fish fossils in the Niobrara chalk in this general area.

The erosive power of waves has exposed good outcrops along the shore of Harlan County Reservoir. Along the south shore at the east end of the lake in the Cedar Springs Campground area, you can see outcrops of Cretaceous Pierre shale, overlying gravels, and some mantling loess. Because the Cretaceous units dip very gently to the west in this area, the Pierre shale is exposed, rather than the Niobrara chalk, which is exposed just east of here. Black shales of the Pierre formation represent continued deposition in the Western Interior Seaway. Veins of gypsum cut across the strata here and are common in the Pierre shale in general. In the 1980s the Kansas University Museum of Natural History excavated a 30-foot-long mosasaur from along the south shore.

An outcrop of white weathered chalks along a gravel road on the south side of the Republican River valley south of Bloomington.

Overlying the shales are tens of feet of gravels that include pebbles of crystalline rocks from distant sources as well as water-worn pieces of the local Cretaceous bedrock. Some of the pebbles are vesicular and porphyritic volcanic rocks that probably originated in volcanic fields exposed in central Colorado. The volcanic rocks are either vesicular or porphyritic—visible crystals are imbedded in a fine-grained matrix. An ancestral South Platte River draining the Rockies may have deposited these pebbles. The distinctive anorthosites from the Laramie Mountains in Wyoming seem to be missing at this location.

The basal sands and gravels were deposited in well-developed crossbeds with trough-shaped cross sections. This kind of sediment is typical of deposits at the bottom of river channels. Farther up in the outcrop, the gravels disappear and the sands get finer and finer. The lateral migration of a channel can form fining-upward sequences; finer sediments of the floodplain cover the coarse sediments of the channel bottom. It is not clear whether these gravels belong to the Broadwater formation or the Ogallala group. Loess caps the section in places.

The Harlan County Dam was built in 1952 to control floods. A devastating flood along the Republican River on May 30, 1935, killed more than one hundred people. Water covered the floodplain for a width of 1 to 4 miles, and walls of water 3 to 8 feet high swept down portions of the river valley during the first pulse of floodwater. Little in the everyday appearance of this river indicates it is capable of such behavior.

Looking downstream from the dam, you can see the beginning of the Franklin irrigation canal. This irrigation system allows more intensive agricultural development in the area. However, at times there is not enough water to meet the agricultural demands and still maintain river flows needed for natural habitat and ecosystems. The struggle to balance competing needs is becoming increasingly difficult for this and many other rivers.

Because it flows along and crosses the state line, the Republican River is also the focus of an ongoing (at the time of this writing) legal battle between Kansas and Nebraska. Kansas claims that center-pivot irrigation systems, which pump water from the shallow alluvial aquifer and spray it over the crops, draw too much water from the Republican River, thus violating water compacts that limit Nebraska's share of the river flow. Nebraska claims that the compacts only apply to surface water and that groundwater is a separate issue. There is no question that the water in shallow aquifers and in the river are connected and that changing one can change the other. But, how much is groundwater extraction affecting the discharge of the river? The answer to this question may help resolve the legal problem. Chances are good that this lawsuit will not have been settled by the time you read this book.

At the base of the Harlan County Dam, the water is released in a large, agitated boil. This release mechanism allows more of the water from the reservoir to come in contact with the atmo-sphere, releasing poisonous gases and adding oxygen. Waters in reservoirs often become stratified, with a stable layer of warm water at the top and isolated cold water at the bottom. The cold water becomes oxygen-poor and enriched in carbon dioxide as organic material decays. If the oxygen content gets low enough, anaerobic decay enriches the water with hydrogen sulfide, a poisonous gas. These waters can be toxic to fish and other organisms downstream.

The road from Harlan County Reservoir to Arapahoe travels mainly along the modern floodplain.

U.S. 275
Omaha—U.S. 20
189 miles

U.S. 275 follows the Elkhorn River for much of its length. Those traveling it will come to know the river quite well and see the changes that occur along its length. It is also an opportunity to become acquainted with environmental issues associated with groundwater and surface water.

U.S. 275 skirts the fringe of the Omaha metropolitan area to the south and heads west out of town through rapidly developing suburbs. The road drops abruptly over loess bluffs and crosses the Elkhorn River. For several miles upstream and downstream from this point, the Elkhorn River flows right against the eastern wall of the valley, where steep bluffs expose bedrock and sediments. The Dakota sandstone is sometimes exposed at the base of the bluffs. About a half mile downstream from the U.S. 275 bridge and easily accessible by canoe are cliffs of Dakota sandstone with abundant, large crossbeds and horizons of ironstone concretions. Above the sandstone, and sometimes right down to the water, you can see at least two different pre-Illinoian tills; a lower, dark-colored till, and an upper, lighter-colored till. An assortment of loess units has filled in and muted the considerable topographic relief eroded into the tills.

On the west side of the river, U.S. 275 travels on a broad, level surface that is the combined floodplain of the Elkhorn River and the Platte River, the latter of which parallels the highway just a few miles to the west. Where U.S. 275 takes a right angle bend to

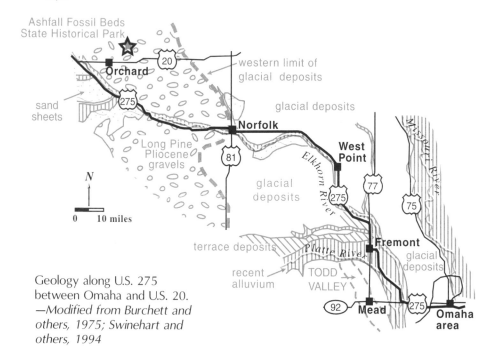

Geology along U.S. 275
between Omaha and U.S. 20.
—*Modified from Burchett and
others, 1975; Swinehart and
others, 1994*

head north, Nebraska 92 continues west across the Platte River and on to Mead. Between the Platte and Mead, the road climbs some moderately steep hills, but Mead is on the edge of a broad, flat geologic feature called the Todd Valley. The scale of this feature is so large that it is difficult to recognize when you are within it. On topographic maps and air photo and satellite images, it appears as a flat, northwest-southeast oriented strip about 5 miles wide between areas of moderately hilly terrain. The valley is higher than, but opens into, the Platte River valley at both the north and south ends and appears to have been the valley of the Platte River at some point in Quaternary time.

Geologists have proposed two theories for the redirection of the Platte River into the Todd Valley. According to the first theory, a minor tributary of the Elkhorn River may have cut back into the landscape by headward erosion and captured the flow of the Platte. The second theory proposes that during Pleistocene time, the last major glacial advance in North America may have redirected Missouri River flow into the Elkhorn River in northeastern Nebraska. Because of its much greater sediment load, the Elkhorn River may have built up a delta where it entered the Platte. The

Bluff exposures on the south side of the Platte River just south of Fremont. At river level is the dark, older pre-Illinoian till. The distinctly dark band midway up the bluff is an ancient soil developed in a younger, pre-Illinoian till. Directly above this is a very thin, dark horizon, an ancient soil in the Gilman Canyon loess. Above it is Peoria loess with a columnar structure.

growing delta dammed the Platte River, causing it to forge a new path down the Todd Valley. When Elkhorn River discharge diminished after the glaciers receded, the Platte River eventually eroded a channel through the obstruction and returned to its previous course, abandoning the Todd Valley.

Peoria loess mantles the Todd Valley, which is underlain by alluvium that follows a classic fining-upward sequence—coarse sediments at the base and finer sediments on top. The alluvium makes a good aquifer. Just south of Mead is a World War II munitions plant that is the source of a large plume of contamination in the groundwater, which flows southeast along the axis of the valley. The plume stretches all the way to where the Todd Valley intersects the modern Platte River floodplain, some 5 miles. Now a Superfund site, contaminated soil has already been excavated and incinerated, and wells have been drilled in the path of the plume to intercept, extract, and clean contaminated water.

Farther northwest, still on the Platte-Elkhorn floodplain, U.S. 275 passes Fremont, joining U.S. 77. Large bluffs rise above the south side of the Platte River just south of town, and just west of where U.S. 77 crosses the river. They are best seen from Hormel Park on the north bank of the river. Pre-Illinoian tills of at least three different ages are exposed here, though only two are obvious to the eye.

The Elkhorn River between Omaha and Norfolk shows the classic features of a meandering stream, including a wide floodplain, tight meanders with point bar deposits on the inside of bends, and abundant oxbow lakes and old meander scars. As the river continues its wandering ways, it further erodes cutbanks on the outside of the meander bends. Some channel portions have been straightened. Riprap along some banks stabilizes them and slows the erosion that consumes adjacent farm fields. For the most part, the valley margins are muted, and the tributary drainages are well developed and vegetated. The river has cut its valley into loess and glacial deposits.

West of West Point, and at other points along the way, you can see feedlots perched on the sloping valley walls. For the feedlot operator, these are advantageous sites; they have natural drainage, are out of the path of any flooding, and are close to a major transportation route. However, sandy alluvium forms a shallow aquifer beneath the floodplain just downhill from the feedlots. Effluent from the feedlot, rich in ammonia that produces nitrates, can contaminate surface and groundwater. Nitrates are common pollutants in Nebraska. Other contaminants, such as hormones given to livestock in their feed, also make their way into the surface and subsurface water. A careful look reveals drainage ditches and retention ponds meant to control feedlot effluents. The Elkhorn has been identified as a compromised water body with contaminant loads greater than the maximum permissible daily loads established by the Environmental Protection Agency. Researchers continue to look into groundwater conditions, bioremediation, and natural reduction of feedlot contamination.

Norfolk is at the western limit of glacial deposits. West of Norfolk, U.S. 275 crosses Long Pine gravels of Pliocene age. Near Neligh are some of the easternmost sand sheets of the Sand Hills. A short distance north of U.S. 275 and Neligh is Ashfall Fossil Beds State Historical Park, a world-class paleontological site.

Ashfall Fossil Beds State Historical Park

Ashfall Fossil Beds State Historical Park is about 8 miles north of U.S. 20, which intersects U.S. 275 east of O'Neill. You can also reach it by heading north on Nebraska 14 at Neligh. The signed gravel road that leads to the state historical park from U.S. 20 dips repeatedly into small valleys that dissect the uplands between the Elkhorn River and the Missouri River. Along draws in the upper reaches of one such valley, the Miocene sediments of the Ash Hollow formation of the Ogallala group have been exposed. Up one of these draws is the rich fossil quarry that is the park. The park is open from Memorial Day to Labor Day.

Fossil mammals from the Ash Hollow formation here and elsewhere in Nebraska have contributed to our knowledge of the diverse fauna that inhabited the plains of the midcontinent about 10 million years ago. The Ashfall site is remarkable among numerous Ogallala fossil finds because of the method of preservation. Volcanic ash blown in from the west filled a watering hole on the plains, entombing numerous animals. Whole herds of short-limbed, barrel-bodied rhinoceroses are preserved in the extensive layer of gray ash along with three-toed horses and other smaller animals, including birds. Many skeletons are complete, especially among the rhinos, posed just as they were at the moment of their death. Paleontologists found small rhino calves with adults presumed to be their parents, and they even discovered pregnant females with unborn young. This unique site not only indicates the diversity of the wildlife at that time but also provides insights into the ecology of that age and the daily lives of the animals.

Mike Voorhies *(left)* working at Ashfall Fossil Beds State Historical Park. The sediments are removed according to a grid system to document the position of fossils. Trenches cut through the ash bed to determine the extent of the fossil deposit form the larger grid in the background.

Mike Voorhies, a native Nebraskan who grew up not far from here and who is now a curator of paleontology at the University of Nebraska State Museum, discovered the site in 1971 when he found a fossil baby rhino eroding out of the gully wall. Museum staff excavated the site in the 1970s; many specimens are now housed in the museum. Excavation was renewed in 1991 after the area was designated a state historical park. The park erected a structure to protect exposed skeletons that have been left in place as a unique exhibit. During summer the excavation continues, and you can see a number of remarkable specimens in place in the shelter of the "Rhino Barn," named for the large number of exceptionally well-preserved rhinos. A small visitor center displays some specimens and sells publications on the geology and wildlife of Nebraska.

Adult and calf rhinos lying head-to-head in the rhino barn at Ashfall Fossil Beds State Historical Park.

A fossil horse in place in the sediment exposed in the rhino barn at Ashfall Fossil Beds State Historical Park.

Ten million years ago, Nebraska's climate was not very different than it is today, and the vegetation, a vast grassland with scattered woodlands, was similar to that of the present day, at least before human settlement. A diverse fauna of plains animals included horses, rhinoceroses, and camels as well as archaic proboscideans. In this respect, the area was more like the veldt in Africa than like the Great Plains, at least in historic times. A crowned crane, a bird that lives in Africa today, has been found at Ashfall Fossil Beds, adding to this sense of similarity.

The remarkable concentration of fossils in the ash bed is undoubtedly the result of the ashfall event. An eruption or series of eruptions from a source to the west, probably in Idaho, produced a tremendous volume of volcanic ash. The ash was blown high into the atmosphere and drifted hundreds of miles downwind before descending on this area in a deadly cloud that blanketed the countryside with fine abrasive material. The herds of rhinos, horses, and other animals that ordinarily used this wet depression as a waterhole may have been forced to stay here as streams and smaller ponds elsewhere dwindled and became choked with ash. They congregated at this last resource but finally succumbed to starvation and thirst. Winds quickly blew in more ash and covered up most of the carcasses. More than 6 feet of ash accumulated.

The ash may have killed the animals in more direct ways as well. A number of rhino skeletons have a peculiar porous encrustation on the bones. This material formed within the bodies of the rhinos in the last days of their lives. In modern animals, this same condition is associated with respiratory disease. While foraging for food in vegetation covered with fine ash, these herbivores continually stirred up clouds of the noxious material and ingested and inhaled large quantities of ash. Volcanic ash, composed primarily of tiny shards of volcanic glass with sharp edges, clogs the lungs and creates numerous, tiny internal wounds. The animals lived long enough for the diseased bone deposits to form.

U.S. 281
Red Cloud—Spencer
208 miles

In contrast to most other routes in this book, U.S. 281 does not follow a natural pathway along a river or travel up the Gangplank surface. Instead, it is a fairly straight north-south traverse that crosses the major river systems of Nebraska—the Republican, Platte, Middle Loup, North Loup, Cedar, Elkhorn, and Niobrara Rivers. The incision of each river valley provides the traveler with a window into the local geology. Terraces and alluvium speak of the recent cut-and-fill episodes of the river system, while older underlying rocks and deposits are exposed in steeper valley walls or in cutbanks. Between the rivers are higher surfaces associated with Quaternary deposits that are mainly vegetated or farmed. Occasional roadcuts, quarries, or small-scale dissection provide a glimpse

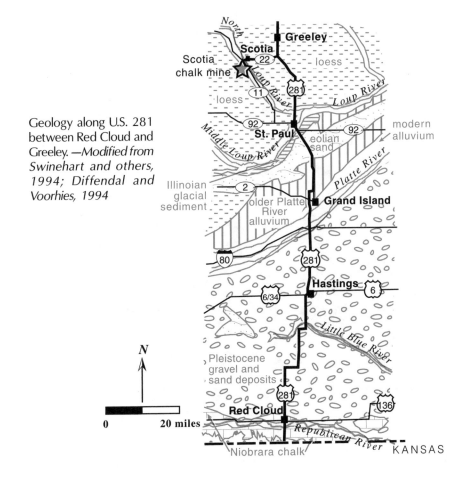

Geology along U.S. 281 between Red Cloud and Greeley. —*Modified from Swinehart and others, 1994; Diffendal and Voorhies, 1994*

into the underpinnings of the landscape. Much of Nebraska's geologic history, from the Cretaceous to modern times, is well represented on this route.

Where U.S. 281 crosses the southern valley wall of the Republican River, there are good exposures of loess deposits more than 15 feet thick. Outcrops of the Niobrara chalk of late Cretaceous age occur at higher elevations to the south, suggesting that these Pleistocene loesses were deposited in a Republican River valley that was a bit wider than it is today. The light coloration suggests that this is Peoria loess, which is associated with the Wisconsinan ice advance. A site up the river valley near Naponee has a terrace deposit mantled with about 30 feet of Peoria loess.

The loess deposits in the valley are thicker than those mantling the higher surfaces on either side of the valley. Why? There are several hypotheses. In the more arid climate of the Wisconsinan

View to the north of a gravel pit wall about 3 miles north of Red Cloud. Note the large crossbeds. Large clasts of local derivation are visible at lower left. The jumbled material overlying gravels in the upper left is recent slope debris that moved downslope from above.

ice advance, more floodplain deposits were probably exposed to wind erosion, so the river valley itself may have acted as a source of loess. Or perhaps, shallow groundwater or occasional flooding in the valley may have supported grasslands that trapped loess. Finally, the valley topography may have provided shelter from the wind, allowing more windblown silt to settle.

Red Cloud, home to author Willa Cather, sits on the edge of the floodplain. Just east of U.S. 281, on the north side of the Republican River, is a rectangular, bermed sewage lagoon for Red Cloud. At times the water level in the lagoon is higher than the road level. The lagoon's elevation helps protect it from flooding and provides a pressure head that helps drive water down into the underlying alluvium. Alluvium often acts as a natural filter for organic wastes, and alluvium filtration increases water quality. The lagoon reminds the viewer of considerations in dealing with sewage treatment along rivers. We discuss the history of flooding along the Republican River in the road guide for U.S. 136.

Loess deposits with subvertical faces overlie sloping gravels in a gravel pit north of Red Cloud. Light-colored Peoria loess overlies the darker Loveland loess, and ancient soils form dark bands within it.

Where U.S. 281 climbs out of the river valley to the north, you can see fairly abundant but small outcrops of the Niobrara chalk. Approximately 3 miles north of Red Cloud, a gravel pit sits on the west side of the road. The Pleistocene gravels contain abundant pink feldspar grains, as well as carbonate clasts likely derived from the Ogallala mortar beds. The types of pebbles in the gravel suggest a Wyoming source. Large crossbeds and channels are nicely displayed in the wall of the gravel pit and indicate vigorous flow conditions with the prevailing current direction to the east. Gravels are more common near the top of the exposed section, and sands are more common lower down. The upper gravels may have been deposited by an ancient Platte River during the Illinoian ice advance about 200,000 years ago. While the Illinoian ice never reached Nebraska, climatic conditions were colder and wetter, influencing the river dynamics. The underlying sand is likely associated with pre-Illinoian ice lobes that initially redirected the Platte River drainage to the south. Although not evident here, to the east some pre-Illinoian lake deposits formed when meltwater became dammed by till or ice.

Above the gravels is 6 to 8 feet of loess with its characteristic vertical columnar fractures and steep profile. A lower section of loess is a distinctly dark reddish brown, typical of the Loveland loess, while the upper section is more yellowish, typical of the Peoria loess. Distinct darker bands represent ancient soil horizons. The boundary between the older Loveland loess and younger Peoria loess is the Sangamon soil, which represents a significant period of nondeposition. The multiple soil horizons within the loess suggest this is an area where deposition rates were much lower than in other places. In general, the loess in this area is only around 10 feet thick. Loess thickness increases as you approach the Sand Hills, especially on the north side of the Platte River. The loess section here is a condensed version of that visible to the north.

Farther north, U.S. 281 ascends to a flat surface. In places, small gullies and steps expose a thin mantle of loess visible from the road. The valleys here have a very striking north-south orientation, suggesting a different underlying fracture pattern than exists elsewhere in Nebraska. Along the 4-mile section of U.S. 281 that jogs east-west, you can look down the length of some of these

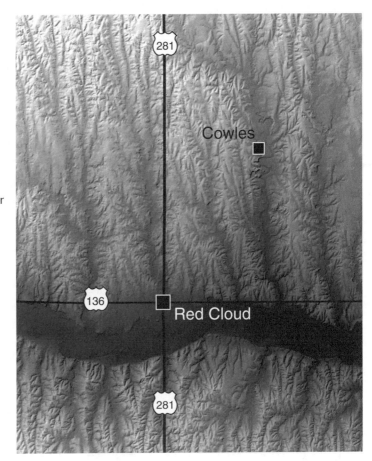

This shaded-relief map shows the very linear north-south drainage pattern controlled by underlying fractures in the Red Cloud region. The image was computer generated from a 30-meter digital elevation model, with two-fold vertical exaggeration, and covers four U.S. Geological Survey 7.5-minute quadrangles (Amboy, Cowles, Red Cloud, Red Cloud Northeast).

drainages. The loess has just enough strength to fracture. Some loess outcrops show a vertical fracture pattern that some geologists have described as columnar. Small-scale erosion processes can concentrate along and follow such fractures, and this influences the direction that gullies and then streams grow.

Underlying the loess are gravels similar to those at the gravel pit north of Red Cloud, and below the gravels in a patch south of the Little Blue River are some Ogalalla group strata, which are underlain in turn by the Niobrara chalk. Outcrops are extremely scarce, and this stratigraphy is known primarily from excavations and drill hole data. The Little Blue River provides the one notable topographic break between the Republican and Platte River valleys. A river terrace runs along on the north side of the valley.

The southern wall of the Platte River valley is a subtle topographic break. U.S 281 crosses four active Platte River channels, two south and two north of I-80. A century ago, even more channels were active, more sand was exposed, and the river was even more braided than today. The route also passes several miles east and downstream of the groundwater gradient of the Cornhusker Ordnance Site. Chemicals from the manufacture of explosives during World War II contaminated the groundwater, and it is now a Superfund cleanup site. We discuss it in more detail in the Nebraska 2 road guide. North of Grand Island, the very flat scenery gradually changes until you are in relatively small, but recognizable, dunes within an elongate dune field just south of the junction of the North and Middle Loup Rivers. The position of the dunes downwind of the Loup Rivers suggests that these broad sandy floodplains were a source of the windblown sand. The town of St. Paul is on the floodplain just upstream from the junction of the Middle and North Loup Rivers.

The North Loup State Recreation Area on the north bank of the North Loup River provides access to the river. A green tower with an antenna on the south side of the river east of the U.S. 281 bridge is a U.S. Geological Survey monitoring station. Springs and groundwater seepage out of the Sand Hills provide much of the flow in the river so discharge fluctuates less than similar sized rivers.

North of the North Loup River, U.S. 281 climbs briefly onto a terrace before ascending into the loess hills. The loess is much thicker here than it is south of the Loup Rivers, starting out at some 15 feet and growing to 50 feet nearer the Sand Hills. A small borrow pit at milepost 100 provides a good exposure of yellowish, stratified loess with laminations. Usually loess is homogenous due to the bioturbation, root growth, and other processes that mix the sediment up beneath the grass cover. The layers in this stratified loess formed either when loess was washed down and deposited at the toe of a slope, or when waterlogged loess slowly slid downslope, especially during Ice Age summers when the surface layer thawed but lower layers remained frozen. About 10 miles north of St. Paul, U.S. 281 climbs up to the upper accumulation surface of the loess. Farther north and in other adjacent areas, fluvial dissection is so intense that only erosional drainage divides remain and the upper surface has been destroyed.

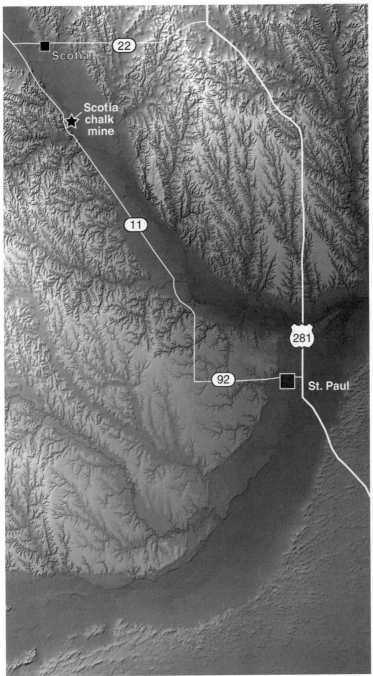

This shaded-relief map shows the junction of the Middle Loup and North Loup Rivers. Note the intricate drainage pattern in the loess, remnants of the upper accumulation surface of the loess, and terraces along the Loup Rivers. Sand dune forms are visible to the south of the rivers. This digitized elevation model was constructed from twelve U.S. Geological Survey 7.5-minute quadrangles (Scotia in the northwest corner and St. Paul Southwest in the southeast corner).

Scotia Chalk Mine

Like many of the other geolocalities in Nebraska, the Scotia chalk mine is off the beaten track. To get there, either head west on Nebraska 22 and turn south on Nebraska 11 or take Nebraska 92 at St. Paul and follow Nebraska 11 along the North Loup River. The geologic attraction is a deposit of impure chalk up to 8 feet thick. The exposures are easily found in erosional bluffs and roadcuts on the edge of the North Loup River floodplain, 1.5 miles almost due south from Scotia. A wayside area with picnic tables, shade, water, and some explanatory signs mark the locality.

The chalk is not the same age as the Cretaceous Niobrara chalk exposed at other places in Nebraska but is of Pliocene age, between 2 and 7 million years old. It is part of the Ogallala group, which extends in the subsurface to a depth of about 300 feet below the chalk layer. Small clam and snail fossils are common, and you can also find burrows and root casts, sometimes filled with fine sand. Some layers are highly siliceous, and diatoms—small single-cell organisms with siliceous tests—make up to 50 percent of the material. The top of the chalk layer gradually changes to thin-bedded, fine-grained sands with abundant rootlets.

Snail fossils on a bedding plank in the chalk at Scotia chalk mine.

All chalk is deposited in standing water—chalk is so fine that if the water were moving, the particles would remain in suspension. Given the fluvial, nonmarine character of the Ogallala deposits, the chalk was probably deposited in an oxbow lake. We don't know the three-dimensional character of the chalk layers, but they are absent elsewhere in the immediate area, suggesting they are not widespread. Similar chalk deposits, also of limited extent, exist elsewhere in the Ogallala sediments. An oxbow lake is shallow and surrounded by wetlands. The root casts and common snail fossils are consistent with a wetlands origin. Since oxbow lakes form when a meandering river cuts off a meander bend, we can infer a meandering river used to flow through here.

In order for such a thickness of chalk to accumulate, the lake had to exist for some time. Such persistence is a bit puzzling, given the relatively dry environment of Miocene and Pliocene time. One possibility is that persistent groundwater springs fed this area. In any case, like the modern groundwater-fed meadows of the Sand Hills, this site was an anomaly—a microenvironment—in Ogallala times and was not necessarily reflective of regional conditions.

Unconformably overlying the Ogallala chalks and sands are Loveland and Peoria loess deposits with a thickness of some 100 feet.

View looking up the North Loup River from Happy Jack Peak. Sandbars and braided channels are evident. Note the town of Scotia in the distance and loess hills on the edge of the floodplain in the far background.

Closed entrance to a chalk mine at Scotia. Picnic grounds at right.

Loess gully north of Nebraska 22 on the way to Scotia from U.S. 281.

Coloration is the key to distinquishing between the two loesses, with reddish brown tones characterizing the Loveland and lighter yellow tones the Peoria. The loess deposits mantle a buried landscape.

A trail climbs to the top of Happy Jack Peak, a mound of loess, from which you can view the modern North Loup River some 160 feet below. A braided channel crisscrosses a wide, vegetated floodplain, and sandbars are common. Test holes indicate the associated floodplain deposits are up to 20 feet thick below the present river bed and represents multiple cut-and-fill cycles.

Happy Jack Peak is thought to have been named after an early trapper and scout, Jack Swearengen, who may have excavated a dwelling in the chalk. The relative softness of the chalk combined with its cohesive strength allow such tunneling. It was subsequently mined commercially starting in 1877 and was abandoned in the late 1930s. The chalk was used for foundation stones, some of which are still in place in nearby Scotia, and in making whitewash, cement, polish, and as a poultry feed nutrient. Some was shipped to Omaha. Two entrances to the mine are near the picnic grounds but are closed up with locked doors. In summer, a local historical group that maintains the mines gives tours.

The landscape east of Scotia shows the classic signature of the loess substrate, with small, erosional, steep-sided rills and landscape steps. Flat-bottomed and vegetated valley bottoms along some now inactive drainages point to periods of erosion of steep gully walls, followed by infilling and stabilization. Periods of active gully formation may have been associated with a drier climate that had less stabilizing vegetation and rare but intense downpours. The loess here is somewhat lighter colored than its counterparts to the east. The east to west rain gradient resulted in more arid conditions here and the greater development of calcium carbonate cementation in the loess as it accumulated.

South of Cedar River is the southern edge of the very distinctive landscape of the Sand Hills. Dune forms and blowouts are particularly well developed both north and south of Cedar River.

Alluvial sediments exposed in the cutbanks of the river display dark bands of ancient soil development, another clue to past climate changes in the area. To the north, the dune forms are replaced by thin sand sheets, and there is greater soil development, more trees, and some center-pivot irrigation systems. The Sand Hills merge into the broad, flat Elkhorn floodplain with O'Neill on the northeast side. There is little indication of what lies beneath the surface.

North of O'Neill, U.S. 281 heads across an area of low relief. You can see the Long Pine gravels of Pliocene time that mantle this surface in small, low roadcuts. Where the road descends into the valley of Eagle Creek, a contact between Ogallala sands and gravels and the underlying weathered Pierre shale is well exposed about three-quarters of the way up the valley wall. The gravels contain fossil bone fragments and some distinctive, locally derived green sandstone and white carbonate clasts. The carbonate clasts may be eroded soil carbonates. Crossbeds indicate a stream current direction to the east. These Miocene-age Ogallala group

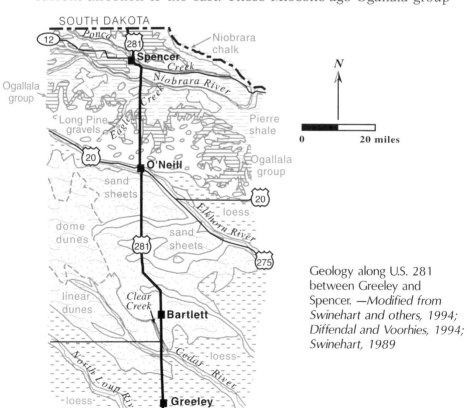

Geology along U.S. 281 between Greeley and Spencer. —*Modified from Swinehart and others, 1994; Diffendal and Voorhies, 1994; Swinehart, 1989*

sediments are very similar to those exposed in the hilltops on the south side of the Lewis and Clark Lake on the Missouri River. You can also see the contact between the Pierre shale and the Ogallala sediments on the north side of the valley at the junction with a gravel road to the west. Seeps mark the contact. The impermeable Pierre shale forms a perfect floor to the High Plains aquifer, the reservoir of subsurface water in the Ogallala sediments. Here, the impermeable floor produces seeps and springs perched high on the valley walls.

The Niobrara River valley provides some excellent exposures of the Pierre shale and river terrace deposits. Spencer Dam, one of the few dams on the Niobrara, has a small, old hydroelectric facility and a recreation area just downstream on the south side of the river. Downstream on the north side of the river are excellent 60-foot-high cutbank exposures of Cretaceous shales, marls, and chalks. You can see sandbars forming downstream of the rapids and the bridge.

The strata in the cliffs vary from black finely layered shales to white and light gray weathered, more massive marls and chalks. Curved fossil tracks, an inch or two wide where some critter crawled

Wind-rippled blowout in the sand dune deposits on the east side of U.S. 281 north of Cedar River.

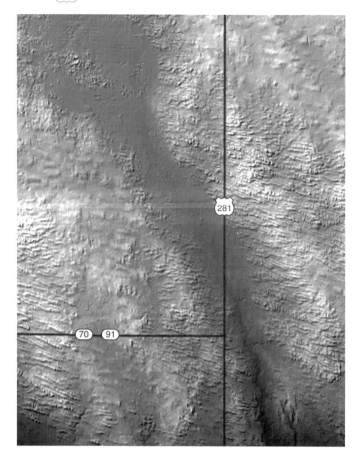

The valley cutting diagonally across this shaded-relief map is Clear River. Note the distinctively different character of this sand dune landscape in contrast to the loess landscape east of Scotia (see map on page 183). A variety of dune forms are evident on either side of the river. Image is a digital elevation model of Bartlett U.S. Geological Survey 7.5-minute quadrangle with two-fold vertical exaggeration to accentuate features.

through the mud, are particularly common in the marls, suggesting that the bottom waters were well oxygenated. You can also find *Inoceramus* shell fragments and carbonized plant remains. The Pierre shale is commonly thought of as primarily consisting of black shales that were deposited in oxygen-poor water. However, the Pierre shale deposited along the eastern margin of the Western Interior Seaway has marls and chalks, formally classified as the Crow Creek and the Gregory members. These units likely reflect some combination of the waxing and waning of the seaway, proximity to the eastern margin, and changing water circulation patterns. Outcrops to the north of the bridge, at a higher stratigraphic position in the formation, display the more typical black shales of the Pierre shale.

A light-colored weathering zone is visible in the cliff at the top of the Cretaceous shales and marls and above this are some 8 feet

View looking upstream from the U.S. 281 bridge over the Niobrara River to Spencer Dam and the hydroelectric facility. The cliff exposures of Pierre shale show a somewhat atypical alternation of black shales and light gray marls and chalks. Above the dashed line are terrace sands with a white weathering zone and an angular truncation of the Pierre strata below. Two arrows show the position of a fault with some 15 feet of offset. A vertical arrow farther to the left shows the position of a small, perched channel cut into the shale and filled with alluvium. In the foreground are sheet metal pilings driven into the river bed to protect the bridge foundations from cutbank migration and erosion.

of brownish, poorly cemented sandstones, which are terrace deposits of the Niobrara River. A lower set of terraces is visible just downstream of the bridge, a narrow one on the north side and its wider equivalent on the south side. Bank exposures upstream of the bridge on the south side show some 5 feet of alluvium on top of the Cretaceous shales. Some small slumps occur along this bank.

A particularly striking feature in the Cretaceous strata at Spencer Dam is a well-exposed set of normal faults. Faults are not common in Nebraska. Two large faults are readily apparent in the north bank. A closer inspection of the cliff shows an array of smaller associated normal faults—faults in which the material above the fault plane moves down relative to the other side. Normal faults occur when the crust is subject to extensional forces. These tectonic

A layer of volcanic ash (labeled with A on either side) shows an offset of about half a foot along a normal fault.

features are related to some minor strain in the interior of the continental plate. They are relatively planar, trend in a generally northerly direction, and have both easterly and westerly dips. Geologists have mapped similar arrays of faults in the Pierre shale in South Dakota just to the north. These faults remind us that the interior of plates are not entirely rigid. The Pleistocene history of glacial loading and unloading caused flexing of the plates and is one of several possible sources of this low-level tectonism. The recent incision of the Niobrara River valley may indicate that tectonic uplift is ongoing in this area. The faults clearly pre-date the overlying terrace deposits, which truncate them and provide some constraint on the time of movement.

Multiple ash layers, centimeters in thickness, are well exposed in the lower slopes of the cliffs. Volcanoes to the west periodically dumped ash into the Western Interior Seaway. Such ash layers are very important for correlating the age of sediments between regions. They can also be particularly weak and are sometimes the surface that a slump slides on.

Climbing north out of the Niobrara River valley, U.S. 281 ascends to an upper, mildly dissected surface, where the road takes a jog to the west before passing through Spencer. This slightly southerly dipping surface is associated with Ogallala deposits just to the west. Just north of Spencer the road drops down again to Ponca Creek, and here Pierre shale dominates the landscape. Just south of the South Dakota border, U.S. 281 climbs up into Ogallala group strata.

U.S. 385
Alliance—Chadron
57 miles

Many Nebraskans drive U.S. 385 on their way to or from the Black Hills. From south to north the road travels through the three major Tertiary groups from youngest to oldest. The Ogallala group is subdivided into several formations that do not extend far beyond this area. They are exposed in the drainage that feeds into the Niobrara River. North of the river, U.S. 385 descends from the Ogallala Tableland, down through Pine Ridge, past Chadron State Park—where Arikaree group strata are particularly well exposed—to Chadron. Jogging west and then continuing north, the road crosses the White River, where Pierre shale is the bedrock, with some patches of White River group strata resting on it.

Though the modern U.S. 385 follows Nebraska 2 for about 10 miles north of Alliance, take the old U.S. 385 if you want to see Carhenge. It is 1 to 2 miles north of Alliance, on the east side of the road. Carhenge, modeled after Stonehenge in England, is one person's offbeat artistic expression that is now a popular tourist attraction; it has the proper astronomical alignments but uses car bodies instead of huge slabs of stone. Unfortunately or fortunately, depending on your perspective, this rust-prone monument will not endure as long as Stonehenge.

Where U.S. 385, the modern highway, descends into the Niobrara River valley, it first crosses Sand Canyon Creek, along which are numerous exposures of the Box Butte formation of the Ogallala group. The type section is exposed upstream from where U.S. 385 crosses the creek. The formation consists of sediment that fills distinctive channels cut into the underlying sediment as

Carhenge in 1993.

much as 140 feet in places. The material in the bottoms of these channels is coarse but derived from the rocks that were exposed in the immediate area at the time and not from the distant Rockies. The finer-grained sediments of the formation consist of mottled greenish to reddish clayey siltstones with white calcareous nodules. Some of the old channels trend east-west, and this section of the Niobrara River follows these channels, perhaps excavating its modern channel in the more-erodable, older valley fill. Fossil quarries in the Box Butte formation have yielded a vertebrate fauna that includes catfish, an owl, various early horses, rhinos, oreodonts, and camels.

Three formations—Runningwater, Box Butte, and Sheep Creek—are older and more localized than the Ash Hollow and Valentine formations of the Ogallala group, but it is convenient to lump them with the Ogallala group sediments, with which they share some similarities of sediment type and depositional style. The Runningwater formation, which is exposed in the area where U.S. 385 crosses Cottonwood Creek and on both sides of the Niobrara River valley, lies beneath and is sandier than the Box Butte formation. Up to five carbonate-cemented ledges in the upper part of

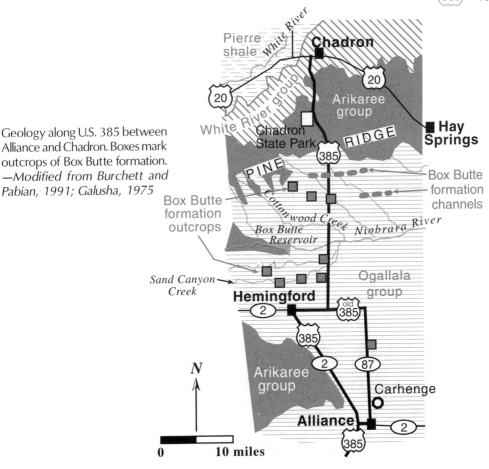

Geology along U.S. 385 between Alliance and Chadron. Boxes mark outcrops of Box Butte formation. —*Modified from Burchett and Pabian, 1991; Galusha, 1975*

the Runningwater formation form a topographic feature known as the Platy Bench. In the vicinity of Box Butte Reservoir, the Runningwater formation lies unconformably on top of the uppermost part of the Arikaree group, equivalent to the Upper Harrison formation. Upper Harrison strata are distinguished from the overlying strata exposed here because they are more massive— composed primarily of sands and silts—and are pink. The Sheep Creek formation overlies the Box Butte formation and has gray and green sandstones, is sandier overall, and is also calcareous.

North of Cottonwood Creek, U.S. 385 crosses the drainage divide between the Niobrara and White Rivers, leaving the mildly dissected Ogallala Tableland and descending the pine-covered canyon walls of Pine Ridge. Outcrops of the Arikaree group are well exposed along the road. On the west side of the road is Chadron State

Park, where prominent bluffs similar to those near Fort Robinson provide a scenic setting. While called Pine Ridge, this feature is more a long set of bluffs and attached ridges that separate a lower landscape with scattered badlands of White River strata from a higher landscape of local tablelands of upper Arikaree and Ogallala sediments. The stratigraphic underpinnings of Pine Ridge are the strata of the Monroe Creek–Harrison formation. The Arikaree formations have a greater lateral continuity than the Ogallala formations, and this continuity produces the extensive Pine Ridge. You can see horizons of the large pipelike concretions of the Monroe Creek formation along the road.

While you can drive to the Black Hills on U.S. 385, we suggest that you take U.S. 20 west to Crawford and then head north on Nebraska 2/71. Along this route, you can visit the Toadstool Geologic Park and Hudson-Meng bone bed geolocalities.

Air photo of U.S. 385 crossing the drainage divide between the Niobrara and White Rivers. The divide crosses the area diagonally from lower left to upper right, separating the Ogallala Tableland *(lower right)* from Pine Ridge *(upper left)*. Pine trees darken the canyons. —*Photo courtesy of U.S. Geological Survey*

Nebraska 2
Grand Island—Alliance
274 miles

Many travelers follow this scenic route across Nebraska to or from the Black Hills of South Dakota. Cutting diagonally across the middle of the state, it is dominated by Quaternary geology, from the loess deposits in the Broken Bow area to the Middle Loup River and finally through the heart of the Sand Hills. Nebraska 2 traverses areas with notably different dune forms and geohydrologic settings, reflecting the varied but structured character of the Sand Hills. Near the western end of this route, the landscape changes noticeably to that of a tableland.

Nebraska 2 heads northwest out of Grand Island within the broad floodplain of the Platte River. The very low-relief surface of alluvial deposits between the Loup and Platte Rivers is essentially a composite floodplain. Four to five miles west of Grand Island and to the south of Nebraska 2 is the Cornhusker Ordnance plant. During World War II, the plant manufactured explosives for the war effort. Wastewater was stored in fifty-six earthen evaporation ponds. Liquid wastes not only evaporated as intended but also seeped into the ground, creating a plume of contaminants in the

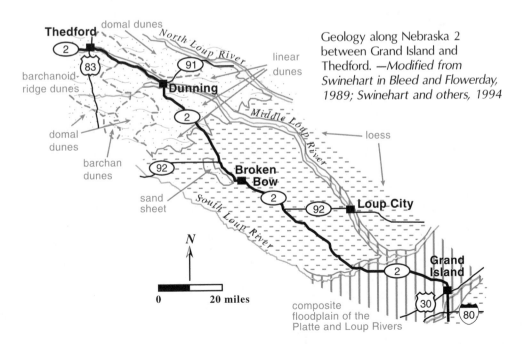

Geology along Nebraska 2 between Grand Island and Thedford. —*Modified from Swinehart in Bleed and Flowerday, 1989; Swinehart and others, 1994*

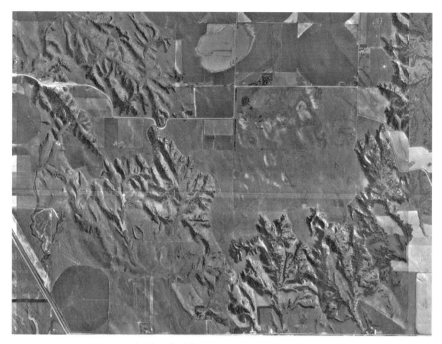

Canyons cut into a tableland of loess just northwest of Broken Bow show a northwest alignment. The road in lower left corner is Nebraska 2. The light patches just off center are likely isolated sand patches on top of the loess. Circular features are center-pivot irrigation fields. Image area is about 4 miles wide. —*Photo courtesy of U.S. Geological Survey*

groundwater that has extended as far as 7 miles from the site. The plume contaminated approximately five hundred private water wells in Grand Island, and the area was designated a Superfund site. By 1988, some 40,000 tons of soil had been excavated from the site and incinerated to remove the source and prevent further contamination. In 1999, a pump-and-treat system was built, whereby the water is pumped out of the ground, cleaned to drinking level standards using a carbon filtration system, and then released to an on-site ditch to sink back into the underlying sediments. In retrospect, a floodplain with a shallow aquifer may not have been the best place to build an ordnance plant, but an awareness of such environmental concerns was not prevalent at the time.

Where Nebraska 2 crosses the South Loup River, geologic maps indicate the bedrock is Pierre shale, but there are no exposures. To the west, the road follows the floodplain of Mud Creek, which dissects loess deposits. The creek drainage has as many as five terraces of Holocene age, some of which are exposed in the Litchfield area. Terrace formation may reflect changes of climate, but very subtle crustal warping may also contribute to the recent downcutting.

Some of the thickest loess deposits in North America occur between Ansley and Broken Bow. You can see loess hills rising above the floodplain of Mud Creek. Thick loess deposits are often associated with grasslands marginal to deserts. The loess here, on the leading edge of the younger Sand Hills, suggests that arid, desertlike conditions may have characterized the middle of Nebraska not only in Holocene time, but well back into Pleistocene time.

Peoria loess mantles the top of the landscape and is up to 150 feet thick. Tablelands represent remnants of the upper surface of loess accumulation before dissection carved the modern topography. In borrow pits several miles due north of Broken Bow and in other locations, the lighter-colored Peoria loess mantles an older topography cut into Gilman Canyon loess and soils. The Gilman Canyon loess mantles still older erosional surfaces. In this way, older, now-buried topography influences the present topography. Periods of erosion and development of topographic relief alternating with periods of loess and alluvium deposition and diminishing topographic relief characterize the geologic history here. Three periods of significant erosion occurred from about 75,000 to 50,000 years ago, from 20,000 to 19,000 years ago, and from 10,000 years ago to present.

A strong northwest-southeast orientation of stream directions is not so easily recognized from the ground level but very evident from higher vantage points or on a topographic map. Sections of Nebraska 2 southeast of Broken Bow follow this trend. Streams and rivers are often oriented in similar directions in jointed hard rocks, but this type of drainage network is less common in unconsolidated Quaternary material. One possible explanation is that a prevailing wind direction that deposited the loess somehow influenced the direction of later stream cutting, possibly by producing landforms or by orienting the sediment itself. A more likely explanation involves loess fractures. Weak cementation in the loess gives it enough strength to fracture; you can see loess fractures in outcrops. Measurements show that one set of these fractures has

View to the west over the edge of the Nebraska National Forest in the Sand Hills. Fires created the patchy distribution of trees.

a regionally consistent, northwest-southeast orientation. Since these deposits are young and never have been deeply buried, very recent near-surface tectonic forces may have fractured the loess.

Just west of Anselmo is the eastern edge of the Sand Hills. The dune type here is linear. The dune forms are relatively low and not clearly delineated because the sand cover is relatively thin. They build into more recognizable dune forms on the east side of Dunning. Between Dunning and Thedford, the road follows the Middle Loup River fairly closely. The river meanders over a broad, sand-dominated floodplain. West of Halsey, some of the dune and interdune sediment is visible from the road where the river is at the edge of its floodplain and cuts into this material.

Near Halsey is the Nebraska National Forest, an interesting anomaly in the heart of the Sand Hills, which are dominated by grasslands. Some 20,000 acres of trees, mostly Ponderosa pines, are nestled between the Middle Loup and Dismal Rivers. This forest was planted as part of a large-scale experiment begun in 1905 at the instigation of University of Nebraska professor Charles Bessy. The purpose of the plantation was to provide timber for development. In addition, some scientists thought the trees might create a microclimate that would increase local rainfall, which in turn would help the forest expand. This arguably was an early attempt to take advantage of a phenomenon now known as the butterfly effect—the tendency of large, complex systems such as atmospheric circulation and weather to be sensitive in the long run to tiny

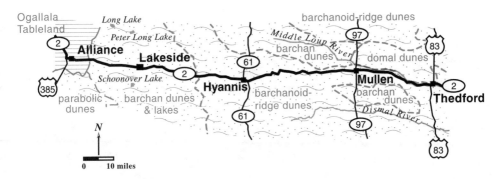

Geology along Nebraska 2 between Thedford and Alliance. —*Modified from Swinehart in Bleed and Flowerday, 1989; Swinehart and others, 1994*

changes. Figuratively speaking, the flapping of a butterfly's wings causes turbulence that may disturb slightly larger air flows that in turn influence still larger currents, and so on, until the result is a hurricane. At Nebraska National Forest, the pines have not successfully reproduced, so the forest must be maintained artificially by planting.

Scientists continue to explore the extent to which variation in the characteristics of the Earth's surface can influence local climate. The Sand Hills have a distinctive surface over a sufficiently large area that this appears to influence atmospheric convection by initiating uplift over the hot surface. The land-atmosphere interaction was probably even stronger when the sands were exposed and actively moving, without any grass cover.

The forest does have one particular effect on the local area—it increases the fire hazard. Lightning ignites the trees, and the forest provides ample fuel to stoke the fire. In 1965, the biggest fire burned about one-third of the forest, about 10,920 acres, along with 7,280 acres of adjacent rangeland. Other fires have consumed hundreds of acres at a time. The nursery at the entrance to the national forest grows seedlings for planting, but there is debate about whether to continue this practice or allow the forest to dwindle. A fire watchtower on one of the higher dunes in the forest is sometimes open to the public, and from here you can get the best view of the Sand Hills possible without an airplane or hot air balloon.

View of the Middle Loupe River valley east of Mullen with rail line in the foreground. Note the blowout in sand dune material in the lower left. Stratified Pliocene sediment underlying the dune sands is exposed in the two cutbanks.

West of Thedford, the dune type changes to large barchanoid-ridge dunes. Roughly upstream from this point, the Middle Loup River is entrenched in a pronounced valley, cutting down into underlying Pliocene deposits, which you can see just east of Mullen. What causes the change in river morphology is not entirely clear, but the Dismal and Snake Rivers are also entrenched where they follow interdune depressions between large barchanoid-ridge dunes. Groundwater feeds the Loup, Dismal, and Snake Rivers. It is possible that as the rivers began to erode their channels, more groundwater was discharged from the large, continuous adjacent dunes, further eroding and entrenching the channels—a feedback loop that perpetuates the entrenchment.

West of Mullen, Nebraska 2 follows flat, linear interdune depressions for long distances. A close look at the steep faces of the dunes on the north side of the road shows that erosion has cut small gullies into them. Small alluvial fans issue from the gullies at the base of the slope. These eroded dune faces are exceptional in the Sand Hills. Elsewhere, rapid infiltration into the sandy soil accommodates all rainfall. Neither surface runoff nor dissection occur. Here, the slopes are so steep and high that during intense rainfalls, water erodes, mobilizes, and redeposits the sand, although only over short distances. The sand moves more as a debris flow—a slurry of water and sand—than as sediment suspended in running water.

These Sand Hills landforms have played a key role in dramatically changing our understanding of a world-famous dinosaur locality half way around the world. In the 1920s, an American Museum of Natural History expedition to the Gobi Desert in Mongolia found the first dinosaur eggs recognized by paleontologists in nests in 100-million-year-old Cretaceous sandstones that were formerly dune sands. Paleontologists believed the nests were those of *Protoceratops,* a small dinosaur with a bony frill at the back of its skull whose fossils are common in this sandstone. In recent years, a new generation of paleontologists from the American Museum of Natural History returned to the same area in Mongolia and made some surprising discoveries. They found nests with skeletons of the parents preserved atop the eggs, apparently brooding them. They identified the parents as the predatory dinosaurs *Oviraptor,* confirmed by tiny fossil embryos in some eggs. This was surprising because paleontologists originally thought these dinosaurs were in this area to feed on the eggs, not raise them.

Paleontologists originally thought that dunes, moving rapidly during sandstorms, overwhelmed the nests, sometimes burying a parent with the eggs. But David Loope, a professor at the University of Nebraska who specializes in eolian processes, suggested that these Cretaceous dunes of the Gobi Desert were much like the dunes of the Sand Hills. The dunes were not actively migrating, and vegetation, though not grasses, covered and stabilized them. However, slurries of sand and water brought down by sudden cloudbursts could have been responsible for the rapid burial and preservation of the dinosaur nests. The base of the relatively sheltered lee slope of large dunes would be a preferred nesting site and possibly close to springs seeping from the sand. In this way and others, understanding the Nebraska Sand Hills has benefited our understanding of other deserts past and present.

Near Lakeside, you can see abundant small lakes on either side of the road. Lakes are particularly common in this part of the Sand Hills. The area extends a bit over 20 miles in an east-west direction and almost 50 miles north-south, with the south end at the

Crescent Lake National Wildlife Refuge. Other areas with similar dense populations of lakes occur, such as the Valentine National Wildlife Refuge, while other parts of the Sand Hills are relatively dry. Why? Ostensibly the groundwater table is shallower and nearer the surface in areas with lakes. One explanation involves dune dams and sand-drowned drainage systems with elevated water tables, and this may very well apply here. The dune sands may have dammed a previously existing drainage and also largely filled the basin with sand, which in turn became saturated with water. Between the dunes, the water appears as lakes. Dune sand thickness may also play a role. The dune form also changes at the edge of the area that is dense with lakes, from the long barchanoid-ridge dunes to widely spaced, crescent-shaped barchan dunes. These barchan dunes form where less sand is available, consistent with their location at the edge of the Sand Hills.

Five to six miles east of Alliance, the rolling Sand Hills change abruptly to a flat tableland. With a mantle of loess on top of older Ogallala group sediments, this surface represents a preserved portion of the regional Gangplank surface.

Nebraska 12
South Sioux City—Valentine
235 miles

While it's not the quickest way to travel east or west, Nebraska 12 is more scenic and exposes a greater diversity of geology and geomorphology than other faster routes, such as U.S. 20. Nebraska 12 crosses or follows the Missouri, Niobrara, Ponca, and Keya Paha Rivers, all of which dissect the landscape and reveal the underlying rocks. From east to west, the road gradually climbs through most of the Cretaceous section and up into the overlying Ogallala group strata. A key unit along this route is the Pierre shale, which figures prominently in geologic concerns relating to both mass-wasting features known as slumps and consideration of this area as a possible storage site for low-level radioactive waste. Near its western end, Nebraska 12 cuts across the Niobrara River valley. First-time visitors to this hilly and scenic part of the state often exclaim that they had no idea there was anything like this in

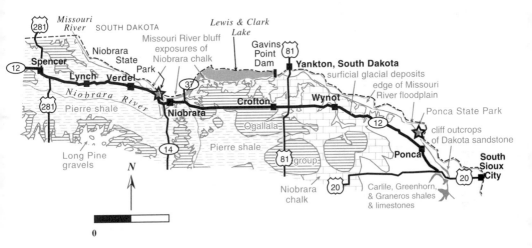

Geology along Nebraska 12 between South Sioux City and Spencer. —*Modified from Diffendal and Voorhies, 1994; Burchett and others, 1988*

Nebraska, and the stretch of river here is very popular with paddlers. Tall cliff exposures of the Ogallala and White River group strata occur along the incised meanders, and vertebrate fossils have been collected from the Ogallala group in this area.

West of where U.S. 20 leaves the Missouri River floodplain, Nebraska 12 splits off to the north and winds through hilly and partly wooded country with few outcrops or exposures. At Ponca, signs direct you to Ponca State Park a few miles north of town on the Missouri River. The Cretaceous Greenhorn limestone and Graneros shale are well exposed there and overlain by the Peoria loess.

Ponca State Park

Ponca State Park, along the Missouri River a few miles north of Ponca, is probably the best locale in Nebraska to see the various Cretaceous units that sit between the Dakota sandstone and the Niobrara chalk. The best cliff exposure is easily accessible at the state park boat ramp and includes a 150-foot section of the Graneros shale and Greenhorn limestone. Other cliff exposures occur for more than a mile downstream, and smaller outcrops occur throughout the park.

In keeping with the western tilt of Cretaceous strata, older and thus lower stratigraphic units occur to the east. The Dakota sandstone, the oldest Cretaceous rock, is exposed at the downstream end of the suite of outcrops.

The units between the beach and river sands of the Dakota group and marine chalk deposits of the Niobrara formation reveal the growth of the Western Interior Seaway. These interbedded sandstones, shales, and limestones are shoreline, coastal, and shallow marine strata. The shifting depositional environment of a growing sea explains the much greater variability of sediment composition here. The contacts between the stratigraphic units are transitional, suggesting that depositional environments shifted in a slow, ordered

Cliff on entrance road to Ponca State Park boat ramp. Blocks of Greenhorn limestone at the base of the cliff fell during the recent mass-wasting event, which took out the young trees. Peoria loess overlies the well-bedded Greenhorn limestone exposed in the middle of the cliff. Graneros shale lies beneath the limestone.

fashion as the sea expanded. However, the overall deepening and widening of the sea was interrupted briefly before the deposition of Niobrara chalk. A period of erosion or nondeposition that lasted for a period of several million years formed a disconformity between the Niobrara chalk and the underlying Carlile shale.

At the boat ramp cliff the more recessive siltstones belong to the Graneros shale, while the overlying more resistant limestones represent about 8 feet of the well-bedded Greenhorn limestone. The overlying Carlile shale is not exposed here, but can be found farther west along Nebraska 12. These relatively thin units are very persistent regionally and are recognized in Kansas, Colorado, South Dakota, and elsewhere. One possible explanation for their widespread occurrence is that sea level changes over a very stable, flat platform have a widespread effect.

Three very thin, yellowish bentonitic clay layers, altered ash horizons, are present in the Graneros shale. Such ash beds serve as crucial marker horizons in Cretaceous stratigraphy. Subduction-related volcanism on the western edge of the North American continent provided the ash. Fossils are uncommon, but those present indicate the Graneros shale is marine. Burrows are common. These were probably well-oxygenated, nearshore, shallow waters where organic materials were mostly decomposed. In keeping with a deepening sea, the siltstones become more calcareous higher in the stratigraphy.

The overlying Greenhorn limestone is notably fossiliferous, and you can find—but not collect—large, particularly well-preserved clams called inoceramids. Be careful if you try to access the Greenhorn limestone in the upper portions of the cliff—it is unstable. Luckily, several large blocks have slid down and are accessible at road level. At the time the limestone was deposited, water depth had increased, and the shoreline had retreated far enough that there was little influx of silt or sand from rivers. Silts reappear in the Carlile shale, which was deposited when the sea receded a little. The Greenhorn limestones represent the deepest water of the

Greenhorn cycle, a cycle of an encroaching and then receding sea. From this time on, biological processes dominated deposition at the bottom of the Western Interior Seaway in Nebraska.

Farther downstream, you can find outcrops of Dakota sandstone. Here, the Dakota is given group status, instead of being a mere formation, and the sands and silts belong to the Omadi sandstone, an upper formation within the group. While the sandstones dominate the outcrops because they are most resistant to erosion, most of the strata are shales and siltstones. Crossbeds are common, but the rocks do not contain fossils. The strata formed in mixed fluvial and shoreline settings. By walking back upstream, you can follow the encroaching sea phase of the Greenhorn cycle.

Small rockfalls and slumps are common along the cliffs. With the Missouri River right against this bank at present, geologically recent cutbank erosion has steepened these cliffs. The shales and bentonites help promote mass wasting. However, this is a different style of mass wasting than the big and complex earth slumps common in the Pierre shale to the west. Here, individual blocks fall, topple, and slide down a much steeper slope; in areas with Pierre shale, entire sections of the hillside slide on a deeper, curved surface.

The bluff line and river here are notably linear in a northwest direction. Joints in the bedrock may control the erosion. A northwest joint set is common throughout Nebraska. A second set of joints with a northeast direction are well developed in the Greenhorn limestone.

Those familiar with the Missouri River in the Omaha area will note a somewhat different appearance here. Multiple channels and sandbars are much more common. The Ponca stretch lacks control structures such as groins. Old air photos from the Omaha area indicate the river there used to have a similar appearance to that of the Ponca area. To protect state park facilities from river erosion, rock groins and riprap were installed in the boat ramp area in 2000.

The topography is very different on either side of the Missouri River—the South Dakota side is flatter than the Nebraska side because South Dakota was glaciated during the most recent ice advance, the Wisconsinan. An ice lobe extended down the eastern half of South Dakota and stopped at what is now the state border. The Missouri River through much of South Dakota followed the western edge of the lobe and then curved east at the southern end of the lobe. This ice lobe essentially set the position of the modern Missouri River.

One mile east of Maskell, the contact of the Carlile shale with the overlying Niobrara chalk is exposed in roadcuts and the cutbanks of streams. Gypsum, of the clear mineral variety called selenite, occurs in veins in the Carlile shale.

Large erratics in fields along Nebraska 12, especially in the higher areas, such as near the junction with U.S. 81, clearly indicate a substantial thickness of older, pre-Illinoian glacial deposits that overlie an erosional surface of Cretaceous strata. Where the road crosses West Bow Creek, there are gravel pits in the floodplain alluvium to the north of the road. It is likely that a significant part of the gravels are reworked glacial deposits. Farther west near Crofton, a straight stretch of Nebraska 12 crosses the top of a tableland capped by a thin layer of Ogallala strata, although exposures are few and small. The contact between the Ogallala sediments and the underlying, weathered Pierre shale is exposed in a roadcut about 13 miles west of Crofton. Farther west, Nebraska 12 curves as it descends to the Missouri River floodplain. You can see a particularly large glacial erratic some 7 feet across about 300 feet south of the road. At lower elevations larger outcrops of the Pierre shale occur.

Nebraska 12 then follows the southern edge of the Missouri River floodplain west to Niobrara. Extensive wetlands have formed at the upstream end of Lewis and Clark Lake, which was impounded by the Gavins Point Dam. Nebraska 121 north of Crofton leads to the dam. Completed in 1957 at a cost of $51 million, the primary purpose of the dam was flood control and power generation. The dam, which is managed by the Army Corps of Engineers, has a visitor center with natural history displays, including information on the local geology. As the farthest downstream dam in a sequence of six dams on the upper Missouri River, this dam plays a particularly important role in the management of the river. The dam has reduced the amount of sediment transported downstream by the river. As of this writing, the Army Corps of Engineers is reviewing and changing a management policy that has been in place for about fifty years. Sandbars and shallow channels are important river habitat. Because the dams trap sediment, and bank control structures eliminate shallow backwaters and channels, wildlife habitat has largely disappeared downstream of the dams for hundreds of miles. One goal of the policy change is to restore some of the habitat.

Just east of Niobrara, Nebraska 37 heads north across a bridge over the Missouri River. On the South Dakota side are some highly visible white cliffs about 100 feet high. Such cliffs intermittently line the Missouri river floodplain on both sides for about 6 miles upstream. The cliffs are Niobrara chalk, which is relatively resistant to erosion and slumping in comparison to the shales above and below it. The type section for this unit is near the town of Niobrara, hence it's name. These strata are Nebraska's and South Dakota's equivalent to the white cliffs of Dover in England. These chalks have a similar origin and reflect unusual global conditions when they were deposited. One of several localities where you can more easily visit outcrops of the Niobrara chalk is in Niobrara State Park. Large fossils that occur in this formation are somewhat

Cliffs of Niobrara chalk on the South Dakota side of the Missouri River immediately upstream of the Nebraska 37 bridge.

uncommon in this area but those found are often well preserved. Fish scales and shark teeth are not uncommon and can be found with diligent searching. The chalk is composed primarily of calcareous platelets produced by planktonic algae.

East of where Nebraska 12 crosses the Niobrara River, Nebraska 14 departs to the south. Just 2 miles south of the junction on Nebraska 14 is a particularly good place to see large slumps that occur to the west of the road. The Niobrara River flows against the base of the slope, undercutting it. The terminology for mass wasting or landslide features is complex and not entirely agreed upon. If the material involved is not hard rock, but unconsolidated sediment or weak and weathered material, and if arcuate scarps and slip surfaces are involved, then *slump* is the most appropriate term.

The largest slump along Nebraska 14 is about 1,000 feet wide. The highway had to be rerouted in 1994 to stable ground, and you can still see remnants of the old road on the slide. Much of the surface slump material is Quaternary sediment, but it is the Pierre shale beneath it that is weak enough to allow the slope to slip. The Niobrara River continually undercuts the toe of the slide, reducing the slope's resistance to movement. The construction of the initial road placed an additional load on the slope and may have altered its internal drainage. Such factors destabilize slopes, and when you add Pierre shale to the mix, a myriad of mass wasting features—in this and many other areas—results. This slump is still periodically active during wet seasons. Knox County, with its underpinnings of Pierre shale, has some forty documented landslides that affect roads. You can see more along Nebraska 12 to the west.

Niobrara State Park

With the Missouri River bounding its north side and the Niobrara River its east side, Niobrara State Park offers a diversity of geology. The oldest exposed rock is the Niobrara chalk in bluffs along the Niobrara and Missouri River floodplains. Above this sits the Pierre shale. Both are fossil-rich legacies from the Western Interior Seaway. In the park you can see slumps in the steeper slopes at the edge of the floodplains and fluvial dynamics associated with damming the Missouri, including a new delta that grew where the Niobrara River enters the Missouri River, long after the Lewis and

Clark expedition camped at the confluence. Like most rivers, these rivers flood. Not only has the town of Niobrara moved to higher ground because of flooding but so have the state park facilities.

The Niobrara chalk is easily accessible in several places in the park, but especially along the Missouri River Trail, which follows an abandoned railroad bed. Sequinlike fish scales and inoceramid clams are some of the most common fossils. A variety of bony fish fossils and shark teeth have also been found in the chalk. Remember that the state park prohibits fossil collecting. Some horizons with burrows likely represent times when the bottom waters were more oxygenated and thus more hospitable to bottom-dwelling fauna. The color of freshly broken rock is dark gray, but it readily weathers white.

The black shale of the overlying Pierre formation does not form many outcrops because it is erodible and weak. The shale is best exposed in slump scars along the line of Missouri River bluffs, but you can also see it in small outcrops along the park road. Pyrite and gypsum are common minerals in the shale. It is not clear why deposition of the shale began and deposition of the chalks ceased, but there is evidence that the waters became cooler, possibly due to the sea developing a connection with colder arctic waters. Colder water would not favor growth of the calcareous plankton, whose skeletons formed the chalk.

A stunning mosasaur fossil was found during construction of a park road in 1986. Mosasaurs were very much like sea-serpents— a large head filled with daggerlike teeth 3 to 4 inches long rested at the end of a 33-foot-long serpent body with four paddle fins. Only the first quarter of the body was preserved—but that's the spectacular end. The specimen was found in the Mobridge member of the Pierre shale, relatively high in the landscape, and the excavation site is marked in the park. You can see a photo of the mosasaur in the state park office, and the specimen is on display at the University of Nebraska State Museum in Lincoln.

Exposed in some of the landslide exposures along the Missouri River Trail is the unusual Crow Creek member of the Pierre shale.

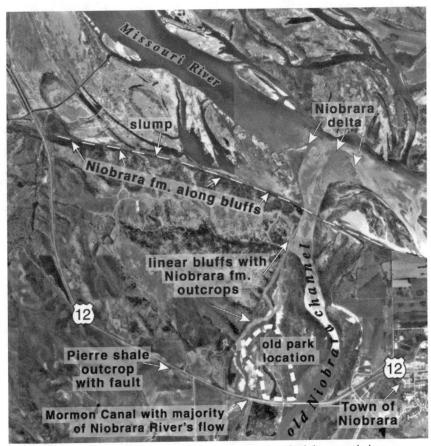

1993 air photo mosaic of the Niobrara State Park, which lies mostly between the Missouri River to the north, the Niobrara River to the east, and Nebraska 12 to the south and west. —*Photo courtesy of U.S. Geological Survey*

It stands out as a light-colored band within the darker shales. The base of the unit is a pebbly sandstone, an unusual deposit for the quiet deep waters of the Western Interior Seaway. An additional oddity is that fossil and rock fragments in the unit are from Paleozoic limestones and the Niobrara formation. Geologists solved the puzzle when they discovered that quartz grains from the sand contain shock lamellae, which geologists believe are formed by meteor impacts. A mega-tsunami caused by such an impact could have created the deposit. The Manson impact structure sits in the shallow subsurface of northeastern Iowa and has been identified as

View looking down the Mormon Canal portion of the Niobrara River from near the RV campground. This portion carries the majority of the flow. White cliffs are exposures of the Niobrara chalk that occur along linear bluffs, the orientation of which is controlled by fractures. Recent deposition of sands and silts and an associated rise in water level partly buried and killed the trees in the floodplain.

the likely culprit. With a 20-mile radius, this is the largest impact structure in the United States and could easily have done the job. Estimates suggest the meteor was 1 to 2 miles in diameter, with a velocity of 45,000 miles per hour, and released more energy than the combined nuclear arsenal of the world. The Crow Creek member thins and disappears to the west, a distribution pattern consistent with a tsunami coming from the east. Early work indicated that the Manson structure formed at the same time as the large impact at the northern Yucatán peninsula in Mexico that caused the mass extinction event at the Mesozoic-Cenozoic boundary, 65 million years ago. More recent work, however, indicated the Manson impact structure is about the same age as the Crow Creek member, 70 to 80 million years old. Given the immensity of geologic time, impact structures are not as rare as many have assumed.

A large earth slump, acres in size, occurs on the north side of the park in the Missouri River bluffs. Walk along the trail on the old railroad bed, and you'll find it where the otherwise level trail climbs up and over a hump of material and bends out towards the river. This hump of material slid down and covered the railroad bed. Given the size of this feature, the rail company should be glad they abandoned this section. While some of the underlying Niobrara chalk moved too, the slump occurs mostly within the overlying Pierre shale. Slumps characterize this region because of the weak Pierre shale, which behaves more like unconsolidated, surficial deposits than bedrock.

This slump is visible in 1993 and 1996 air photos and was active in 2001—displaced, tilted, and even fallen trees still had summer leaves. Large fresh cracks at the hill crest, some well over a foot wide, indicate that the slump will likely cut back farther into the slope and grow in size. Slumps tend to have long and complex histories, with multiple episodes of movement over decades or more. They can reactivate after periods of greater rainfall. In addition, a slump does not move as a single unit. The slump consists of smaller sections that move semi-individually. A new trail was bulldozed into the upper part of the toe of the slope. Ironically, this may have triggered some internal portions to move. Sections of the new trail slid away soon after it was made.

This mass-wasting feature has many of the classic features associated with earth slumps. At the top is a main scarp which exposes the Pierre shale, including the distinctive Crow Creek member. The scarp formed as material along the river slid down-hill. In the middle of the slump are semi-intact slivers of ground some 70 feet wide with trees that once were on top of the hill. The ground surface and the trees are tilted towards the slope. The slip surface that the slump moves along is curved, or scooped, so the material rotated and the trees tilted. Behind these slivers are depressions that collect water. Lubrication by the water can cause the sliver to slump again. At the bottom is the slump's hummocky toe material, which has moved a good 300 feet or more out onto the Missouri River floodplain. The different lobes and the array of dead trees in various states of decay attest to the episodic nature of movement. In the absence of the downstream dam, the toe material would not survive. Left to its own devices the river would once again wander this way and carry this material away.

The position of the slump in the landscape makes geologic sense. The Missouri River used to bend against and cut into these bluffs. The channels have changed since the impoundment of Lewis and Clark Lake by Gavins Point Dam in 1957. Topographic maps show this was one of the steepest and highest slopes along this stretch of river. Its height also meant that more Pierre shale was available to slide. You could expect a slide in this high, young, steep, and weak slope. You might wonder if building the railroad was involved in triggering this feature, but the line runs along the base of the slope rather than cutting back into it. Other smaller slumps also exist along this bluff.

At the Niobrara and Missouri River confluence, a delta extends out into the Missouri River. A fan-shaped collection of sandbars, the delta can be seen easily from some of the trails on top of the bluffs. The delta didn't exist prior to 1938, well before Gavins Point Dam was built downstream of the confluence, but by 1975 it was well established. Periodic floods on the Missouri used to sweep this material away, but with the advent of the dam, such sediment clearing has not occurred. The delta now constricts flow and acts as a partial dam.

Gavins Point Dam also caused the Missouri River level to rise upstream of it. The Niobrara River responded by building up sediment, 7.5 feet in places, in its lower reaches from 1952 to 1996. The rise in the level of the riverbed caused channels to shift and increased problems with flooding. While the majority of the flow used to be in the eastern channel, most of it now moves along the western channel, also known as the Mormon Canal, perhaps originally dug by Mormons to provide power for a mill. A 1960 flood with a peak discharge of 40,000 cubic feet per second took out the Nebraska 12 bridge. Cabins at Niobrara State Park, originally located on the river floodplain, filled with sand. The park abandoned the facilities and moved to higher and more scenic ground, reopening in 1987. An unintended effect of Lewis and Clark Lake was to elevate the Niobrara riverbed, which localized flooding concerns along its lower portion.

A large slump feature in Niobrara State Park along Missouri River bluffs. *Top:* Main scarp. Material to the right slid downhill exposing the Pierre shale. Note the vegetation is disturbed but alive, with trees to the very right tilted back towards the hill. Some of the whitish material exposed in the main slump is the Crow Creek member. *Middle:* Cracks in material bulldozed for a new path. This is the upper part of an internal slump within the larger feature. It turned into a debris flow farther downslope. *Bottom:* The hummocky, lobed toe of the slump that moved out into the Missouri River. Note uprooted and downed trees.

Outcrop of Pierre shale on Nebraska 12 west of the entrance to Niobrara State Park. A fault or an old slump's slip surface causes the approximately 30-degree tilt of the beds to the left.

On Nebraska 12, just west of the entrance to Niobrara State Park, are two large roadcuts through the Pierre shale. Septarian concretions and clear gypsum (selenite) occur in the shale. At the southeastern end of the cut on the north side, the strata dip anamolously at 30 degrees, while farther along in the cut they only dip shallowly. A fault separates the two sections of strata but is difficult to see now because of grass cover. Some debate exists as to whether this is a tectonic fault or a slip surface associated with mass wasting. The inclination of the strata farther back in the hill, the position of this hill on the edge of Niobrara River valley, and the lack of a continuation of the fault in the underlying chalk cliffs all argue for a mass-wasting origin.

About 4 to 5 miles west of Niobrara, Nebraska 12 descends to the Missouri River floodplain where extensive wetlands exist. While not ostensibly part of Lewis and Clark Lake, the wetlands are the result of the hydraulic effects that the reservoir imposes on the

Geology along Nebraska 12 between Spencer and Valentine. —*Modified from Diffendal and Voorhies, 1994; Burchett and Pabian, 1991*

river upstream of the high-water lake level, and of the growth of the Niobrara River delta out into the Missouri River floodplain. Large stands of dead trees mark areas where the groundwater rose upon impoundment, drowning the tree roots and creating wetlands. In fact, the original site of the town of Niobrara lies within this drowned area. Only a small marker remains at the original location. You can see more cliffs of Niobrara chalk on the South Dakota side of the Missouri. The break in slope at the top of the Niobrara chalk marks the contact with the overlying Pierre shale.

Between Verdel and Spencer, Nebraska 12 follows the floodplain of Ponca Creek, which has cut down into the Pierre shale. Capping the hills are Ogallala group sands of Tertiary age, Quaternary loess, and in some places traces of Pliocene Long Pine sand and gravel. East of Lynch is another large slump. A road sign designates it as a slide area, and you can see fresh exposures of the overlying Long Pine formation and sands in the scarp near the crown of the slump. Another slump west of Lynch on Nebraska 12 necessitated moving the road. The old road was closer to the slope and is now largely covered or gone. There was once a large village of Native American earth lodges that were occupied from A.D. 1450 to 1550 near Lynch.

At Spencer, Nebraska 12 leaves Ponca Creek and climbs up onto a tableland topped with Ogallala sediments, but the Pierre shale is not far beneath. Driving through the small town of Butte, you

Waste repository protest signs just north of Nebraska 12 west of Butte. Less legibly, to the left it says "Save our children, save our land, save Boyd County, take a stand."

can see distinctive but faded signs that refer to the "dump" and U.S. Ecology. These signs are reminders of a long and interesting political battle that continues today, and Butte is basically ground zero. As the United States developed nuclear power, the issue of what to do with the radioactive waste arose. Radioactive waste falls into two categories, low-level and high-level. High-level radioactive waste is highly radioactive and has a long half-life. Since it is much more dangerous than low-level waste and will be with us for thousands of years or longer, the federal government has assumed responsibility for it. Efforts to find an acceptable way to store the high-level waste continue. Low-level waste is less radioactive and decays to relatively harmless levels more quickly than high-level waste.

Federal legislation, called the Low Level Radioactive Waste Policy Act of 1980, mandated that states either take care of their own low-level radioactive waste or form state compacts to share responsibility for the waste disposal. A compact commission would analyze and select the best storage site, using a variety of criteria including geologic suitability. Nebraska, Kansas, Oklahoma, Louisiana, and Arkansas formed such a compact. (Nebraska has two nuclear power plants.) After a broad-spectrum study of potential sites, the commission chose a spot in Boyd County, Nebraska, and selected a corporation named U.S. Ecology to develop the waste site.

The larger of the two buttes west of the town of Butte, with Ogallala mortar-bed capstone.

Although some people saw the site as a potential economic boon to the area, strong local opposition to the repository developed.

The commission chose the site in part because it sits on Pierre shale, which is impermeable to water unless fractured. This characteristic makes it potentially a good burial or storage medium because it would isolate the waste from the groundwater. A plan was developed to store the low-level waste material in aboveground vaults. The time line called for the waste site to open in 1993. However, scientific, political, and economic factors continue to complicate the site evaluation: wetlands and a shallow aquifer are perched on the Pierre shale; fracture permeability does exist in Pierre shale elsewhere; the potential site is just across the border from South Dakota, which does not belong to the compact; and a financial scandal in the project administration brought unfavorable attention to the process. The Nebraska Department of Environmental Quality denied approval, citing the possibility that groundwater could penetrate the facilities. This story is now decades long and as of this writing the compact is suing the state of Nebraska. Very large sums of money are involved, and one estimate is that up to 1996, $80 million had been expended or committed. Estimates in 1994 dollars are that the cost of implementation would be in the range of $90 million over 30 years of accepting waste.

Just to the west of Butte, two small but distinctive buttes rise a quarter of a mile south of Nebraska 12. The cap rock is one of the Ogallala mortar beds, an old soil horizon. If you are driving west, this is your introduction to how differential erosion of these resistant beds shapes the landscape, forming benches, ledges, buttes, and other landforms. This landscape is distinctly different from that underlain by the Pierre shale. The Ogallala group is not subdivided into formations on geologic maps of the area, but these sediments are similar to the Ash Hollow formation farther west. These outcrops remind us how far east the apron of debris shed from the Rocky Mountains extended.

Just east of where Nebraska 137 heads north to South Dakota, gravels and coarse sands of the Long Pine formation of Pliocene age are exposed high in the landscape and in roadcuts along Nebraska 12. Clasts in the gravel include a variety of metamorphic and plutonic rock types. In 1970 geologists proposed that these were glacial outwash deposits, with an origin somewhere to the north. Subsequent work on the types of rocks and the age and the channel configurations of these deposits indicates that the source was in Wyoming and north-central Colorado. These sediments probably represent the beginning of the exhumation of the Rocky Mountains. They were deposited in braided channels, the lowest point in the landscape at that time. Their high position in the present landscape means that all of the steep topography you see today, including the modern drainages, has been carved since the Long Pine gravels were deposited—in approximately 2 million years.

Look for the large cutbanks exposing Pierre shale half a mile downstream of where Nebraska 12 crosses the Keya Paha River. The road climbs from the valley back onto the tableland of Ogallala strata where intermittent outcrops are visible along and close to the road. The numerous wetlands in this area are probably fed by small groundwater systems perched on top of the impermeable Pierre shale. Just west of Springview, two wind turbines add to the portfolio of energy sources Nebraska utilizes. The average wind speed across these plains is 18 miles per hour. This also reminds us of wind's potential for eroding, transporting, and depositing sediments. Near the Keya Paha—Cherry County line, you can see small, isolated dunes north of the road.

West of Sparks, follow signs to Smith Falls State Park to see a dramatic part of the Niobrara River valley. The river has cut a deep valley into the Ash Hollow and Valentine formations of the Ogallala group and Rosebud formation of the White River group. The Ash Hollow cap rock member defines the top of the valley, while underlying Valentine formation sands are good aquifers, erode easily, and form local benches. The Rosebud strata, quite different from the overlying Ogallala strata, are more clay and ash rich, more impermeable, and distinctly reddish or tan. Water associated with perched aquifers in the Valentine sediments seeps out along the contact with the Rosebud formation and stains the cliffs below dark red. Smith Falls, fed by springs, originates at this contact and flows into the Niobrara River. At 70 feet, it is the highest waterfall in Nebraska. In addition to the large cutbank cliff exposures of these Tertiary strata, the park boasts terrace deposits of reworked Tertiary material that contain Pleistocene fossils. Geologists don't yet understand the complex story of how this part of the river recently became entrenched.

This part of the Niobrara River valley is biodiverse. As the glaciers retreated at the end of Pleistocene time, the plants that were

Smith Falls cascades over the Rosebud formation at Smith Falls State Park along the Niobrara River.

Cliff section of mostly Valentine formation exposed along the Niobrara River downstream of Smith Falls. The arrow marks the contact between the Rosebud and Valentine formations. Note paddler on river for scale.

adapted to cold climates retreated with them. But in the moister and cooler valley incised by the river, they persisted. Such isolated ecological systems are called refugia. Stands of birch and aspen are the most visible indicators of this phenomenon in the valley. Eastern, western, southern, and northern species of plants and animals mix here, with many at the extreme edge of their ranges.

Smith Falls State Park is a popular place for paddlers to put in or take out of the river. Because of its scientific and recreational value, a 70-mile-long stretch of the Niobrara River east of Valentine was designated a National Scenic River in 1991.

Just east of Valentine, Nebraska 12 descends from the top of the Ash Hollow formation tableland down into the Rosebud sediments and crosses Minnechaduza Creek near the entrance to Fort Niobrara National Wildlife Refuge. Outcrops are easily accessible. Upstream from here, the Niobrara River valley has a different geomorphic character. It is less entrenched with a broader floodplain and has some areas of lake sediments from times when sand dunes blocked the river.

Fort Niobrara National Wildlife Refuge

A few miles east of Valentine, turn south off Nebraska 12 to the Fort Niobrara National Wildlife Refuge. The wildlife maintained at the refuge includes bison, antelope, elk, and other large animals. In 2000, a herd of longhorn cattle was moved from the refuge to Fort Robinson in an old-fashioned cattle drive.

You can see cliffs of Tertiary formations along the Niobrara River valley from several points in and around the refuge. Fort Falls is easily accessible and well worth a side trip along a trail that goes from the overlook parking area down to the river level. At the base of the Fort Falls cliff, a steep, pinkish slope extends up from the water. This siltstone of the Rosebud formation, with diffuse bedding contacts, was deposited by small streams and as windblown loess deposits during late Oligocene time. The age is based on scarce mammal fossils found in it elsewhere. In places old soil horizons are also preserved.

The Rosebud formation is separated from the overlying Ogallala group by a distinctive erosional surface that is easily recognized by the seeps and springs that leak from the boundary. When groundwater percolating down through the overlying Valentine sands encounters the impermeable siltstone of the Rosebud formation, it flows laterally out of the cliff face. Thus, the Rosebud formation forms the floor of the High Plains aquifer, perched above it in the Ogallala sediments. The stream that forms Fort Falls is also mainly fed from leakage of the aquifer out of the Ogallala and younger sediments.

The Valentine formation, which is completely exposed in these cliffs, consists of sands deposited in the valleys of ancient rivers. Look for the many small channels that have been scooped out and then filled in again. The sands have not been significantly cemented so they form gentle slopes rather than cliffs.

Fossil quarries in the Valentine formation near the town of Valentine have produced a rich assortment of fossil vertebrates of middle Miocene time, about 14 to 12 million years ago. The fossils collected from the Valentine formation seem to paint a very complete picture of the wildlife of that time. Mammals include large animals such as elephants, horses, camels, and rhinoceroses; smaller even- and odd-toed hoofed mammals such as tapirs and peccaries; carnivores related to cats, dogs, and weasels; small mammals such as rodents; and other species belonging to each of the groups listed above but without familiar names because they are only distantly related to any of the mammals living today. Turtles, lizards, alligators, and birds also lived in Nebraska in Miocene time. Plant fossils from the Valentine formation nearby complete the picture of a habitat that was wetter and more wooded than the same area is today. A small museum associated with the wildlife refuge visitor center displays some of these fossils.

At the top of the Fort Falls cliff, the cap rock member is the only part of the Ash Hollow formation that has been preserved here. Erosion has stripped away the upper part. The cap rock member is also a sandstone but is much better lithified than the Valentine. One of the many ledge-forming ancient soil horizons in the Ash Hollow formation, it contains abundant root structures.

View of Niobrara River cliffs from the Fort Falls overlook. The arrow marks the contact between the Rosebud and Valentine formations. The dark streaks are seeps.

Geologists don't know why the Niobrara River has incised its way several hundred feet down into bedrock here. Terrace deposits from late Wisconsinan time high in the landscape indicate the incision began about 12,000 years ago and may have been complete by 5,000 years ago. Whatever the reason, the river provides excellent rock exposures and a mecca for canoeists.

Nebraska 29
Scottsbluff—Harrison
56 miles

Nebraska 29 cuts north-south across modern drainages that expose sediments that filled ancient drainages. Follow U.S. 26 west from Scottsbluff to Mitchell and turn north onto Nebraska 29. You'll climb out of the valley of the North Platte River, cross over a drainage divide, and descend to the Niobrara River, which cuts into the upper units of the Arikaree group of mid-Tertiary age. Agate Fossil Beds National Monument preserves fossils in the sediments along the Niobrara River. These same sediments underlie the rolling grassland all the way to Harrison and beyond. A road that heads north from Harrison crosses Pine Ridge.

The North Platte has cut deeply into the silts and mudstones of the White River group of mid-Tertiary time, but as Nebraska 29 climbs north out of the valley and up the drainage of Dry Spotted Tail Creek, the increasing elevation corresponds to higher, and thus younger, rocks in the stratigraphic section. The road ascends through the Arikaree group, first through the Gering and Monroe Creek formations in the walls of the North Platte River valley, and then through the Harrison formation, which is extensively exposed in the washes of the upper drainages of the creek.

Still higher in the drainage basin of Dry Spotted Tail Creek, about 17 miles north of Mitchell, much younger sediments of the Ogallala group of Miocene time are exposed in small gullies cut back into the high surface that divides the drainages of the North Platte and Niobrara Rivers. These outcrops appear as bright patches of sediments east of the road. Unlike the White River and Arikaree sediments that are traceable over long distances with little change, the Ogallala sediments filled valleys and thus have limited lateral

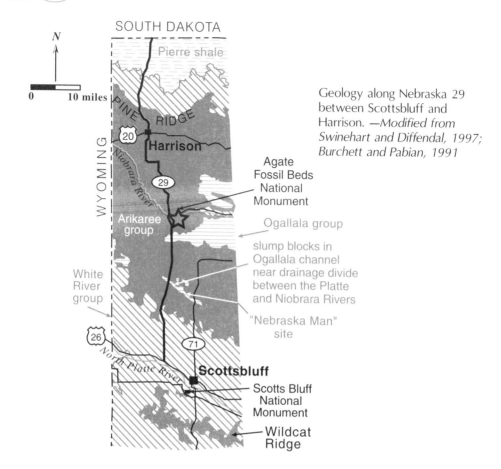

N

0   **10 miles**

SOUTH DAKOTA

Pierre shale

PINE RIDGE

20

Harrison

29

Niobrara River

WYOMING

Arikaree group

White River group

26

North Platte River

71

Scottsbluff

Geology along Nebraska 29 between Scottsbluff and Harrison. —*Modified from Swinehart and Diffendal, 1997; Burchett and Pabian, 1991*

Agate Fossil Beds National Monument

Ogallala group

slump blocks in Ogallala channel near drainage divide between the Platte and Niobrara Rivers

"Nebraska Man" site

Scotts Bluff National Monument

Wildcat Ridge

extent across the valleys. Only deposits that filled the deepest parts of these valleys have been protected from later erosion. Throughout Miocene time, several episodes of erosion cut valleys into older sediments. Each generation of valleys became filled with sediment as the river systems changed from cutting down to filling up. Streams of late Miocene time carved deeply enough to reach the Arikaree sediments in places but also cut into and across the earlier Ogallala valley fills, sometimes obliterating them.

Such complex patterns cannot be understood and interpreted at any one location. It took years of careful fieldwork by geologists and paleontologists to unravel this puzzle. At one outcrop just a short distance from Nebraska 29, sediments of the youngest member of the Snake Creek formation have buried large blocks of the oldest member of the formation. The blocks must have slumped

Exposure of the Snake Creek formation near Nebraska 29. The large, light-colored blocks are from the older member of the formation and slumped into a channel that was completely filled by the youngest member of the formation.

into the channel filled by the younger member. These slump blocks provide the only evidence of the older member at this location.

Nearly a century ago, vertebrate paleontologists collecting for the American Museum of Natural History in New York discovered fossils in these Ogallala sediments. Since then, a large collection of fossil mammals has been produced from quarries in the few square miles of the upper drainage of Dry Spotted Tail Creek and the hills of the major drainage divide.

The Sheep Creek formation and the younger (but not necessarily higher because of the cut-and-fill episodes) Olcott formation were deposited early in Miocene time. The Sheep Creek formation is younger than the Runningwater formation and close to the age of the Box Butte formation that filled valleys farther north near the present valley of the Niobrara River. Based on the fossils it contains, the youngest of Miocene-age formations in this area, the Snake Creek formation, appears to be younger than the Valentine formation discussed elsewhere, and about the same age as the Ash Hollow formation.

The diverse fossil faunas from the Ogallala formations include representatives of several groups of mammals: carnivores such as archaic doglike amphicyonids; even-toed ungulates including camels, ancient relatives of the modern pronghorn antelope, and extinct oreodonts; and odd-toed ungulates such as the extinct chalicotheres, rhinoceroses, and several different kinds of horses. These fossils not only have provided vital clues to help decipher the complex stratigraphy but have been important in understanding the evolutionary history of these mammals as well.

One fossil from this area attracted quite a bit of attention when it was found. Just a few miles east of Nebraska 29, a tooth was collected in 1922 from the Snake Creek formation of the Ogallala group on Olcott Hill along the North Platte–Niobrara River drainage divide. The tooth was interpreted as that of an anthropoid ape that was named *Hesperopithecus*. In the excitement of this discovery, paleontologists attributed additional isolated teeth found in the same area to what they thought was a potential human ancestor—called "Nebraska Man." But only a few years later, when paleontologists were able to look at the evidence with cooler heads, they determined that the original tooth was a worn molar of a fossil peccary, *Prosthenops,* that was common in the Ogallala beds. Fossil humans and human ancestors were being found in Europe, and American paleontologists were eager to find part of the human family tree in their own country. Even though a very worn peccary tooth does resemble a very worn ape tooth, this is not a mistake that they would have made if they had not been so eager to accept their initial conclusions. Creationists like to point to this case as a demonstration of the unreliability of the human fossil record, especially because it seems absurd that scientists couldn't tell a human from a pig. But, it should be noted that the scientific method prevailed and paleontologists detected and corrected their mistake.

Just a couple miles farther north (nearly 20 miles north of Mitchell), the road crosses the divide between the drainage basins of the North Platte and Niobrara Rivers. Though the divide is subtle, it is one of only two major divides in Nebraska. The surface waters the state consist of three major river systems—the Republican, Platte, and Niobrara—that flow from west to east to the Missouri River. Most of the surface of the state drains eventually into the

Platte River with the Niobrara draining the area remaining to the north and the Republican River the area to the south.

North of the divide, Nebraska 29 crosses a low-relief surface developed primarily on Arikaree sediments. But here and there, remnants of the Runningwater, Olcott, and Snake Creek formations of the Ogallala group are preserved where they filled deeply incised valleys. On the north side of the divide, the uppermost formation of the Arikaree group, the Upper Harrison, is sometimes present above the Harrison formation. Although part of the Arikaree group, the Upper Harrison is separated in time from the Harrison below it—a long interval of erosion cut a broad valley into the Harrison that the Upper Harrison later filled. A few miles south of Agate Fossil Beds National Monument, Nebraska 29 drops off this gentle surface and into the valley of the Niobrara River.

The Niobrara has created a broad valley, but the present river, which the road crosses just south of the entrance to Agate Fossil Beds National Monument, seems a mere trickle of water. Perhaps the erosion that shaped the valley occurred during wetter times. The valley exposes the Harrison formation in the lower valley walls and the Upper Harrison above it. Abundant mammal fossils were collected in this area and were used to characterize the faunas of very early Miocene time.

Agate Fossil Beds National Monument

Agate Fossil Beds National Monument protects an area in the Arikaree group that has produced a remarkable assemblage of fossils, which now grace the halls of major natural history museums. The story of the collection of those fossils reveals much about the history of paleontology in America. In the late nineteenth and early twentieth centuries, when this part of Nebraska was still part of the wild frontier, the Cook Ranch at Agate Springs was host to some of the great paleontologists of the time. O. C. Marsh and E. D. Cope, whose professional feud has become legendary in paleontology, both collected here. So did H. F. Osborn of the American Museum of Natural History and E. H. Barbour of the University of Nebraska State Museum. But it was O. A. Peterson of the Carnegie Museum who made the big discovery—a bone bed, a layer of densely packed remains of ancient mammals.

The Cook family led paleontologists to the fossils and supported their operations in the field. The family eventually ended up owning the land that included the fossils sites. Harold Cook became a well-known fossil collector.

A sign on Nebraska 29 marks the entrance to Agate Fossil Beds National Monument. The road to the monument follows the broad valley of the Niobrara River, now only a trickle of water flowing through the middle of the valley. The valley walls expose the younger part of the Arikaree group, the Harrison and Upper Harrison formations.

Stop at a parking lot on the north side of the entrance road, near the junction with Nebraska 29, to see a good example of a *daemonhelix*, a devil's corkscrew, in the cliff wall of the Harrison formation. Interpretive signs along a path discuss these strange features, which are the burrows of the Miocene rodent *Paleocastor*. The spiral-shaped burrows, which may be up to 10 feet long, were a mystery until fossils of the rodent were found within several burrows. You can also find tooth marks on burrow walls. Why the corkscrew shape? Well, it's a lot easier to descend or, especially,

View from a small ledge of an old calcified soil above the excavated den site of doglike mammals in the left foreground. Carnegie and University Hills are in the middle distance with the Niobrara River valley in the far background.

ascend an incline than it is to climb a vertical shaft, and a helical path is the easiest way to go nearly straight down to a significant depth. It is unlikely that these rodents would have dug such deep burrows if flooding were likely, so the water table must have been fairly deep and the climate fairly dry. The lowest outcrops on the east of the trail expose well-sorted sands with large crossbeds. These Arikaree sediments are former sand dunes.

Three miles down the entrance road is the national monument visitor center, which displays fossil specimens and has exhibits that explain the history of the region and the fossil site. The story of the Cook family is informative about life in this area around the turn of the nineteenth century. A trail from the visitor center leads south across the river and uphill to two prominent, conical hills—Carnegie Hill and University Hill for the Carnegie Museum and the University of Nebraska—landmarks well known to vertebrate paleontologists for their remarkable fossils. From the western hill, Carnegie Hill, look to the outcrops to the west for an exposure of a fault that has offset some distinctive ledge-forming beds. The fault parallels the Niobrara River valley on the southern side and has the same general east-northeast orientation as Toadstool fault at Toadstool Geologic Park.

The famous bone beds are in the Upper Harrison formation at the base of each hill. The layer of animal remains accumulated at a waterhole in a streambed that dried up during drought. The bone beds produced an abundance of two-horned rhinoceros fossils, *Menoceras*. They were much smaller than their modern cousins—not even as large as a cow—and the two horns were side by side at the ends of their noses.

The quarries were also rich in bones of the chalicothere *Moropus*. Chalicotheres are an extinct group of odd-toed ungulates related to horses and rhinoceroses but strikingly different in appearance. *Moropus* was as large as a modern horse, but with different proportions—taller but shorter from head to tail. Their long front limbs bore large claws not hooves. Some other ancient large herbivores also had claws, such as giant ground sloths, but there are no modern analogues to help us guess what they used these claws for. *Moropus* had the teeth of a browsing animal, a leaf-eater, so they may have used the claws to grasp small tree branches and pull them down to eat. Or they may have used the claws to defend themselves from predators or to compete for mates. Most likely the claws had multiple uses.

Other types of mammals occur in the bone beds and at other locations in lesser numbers. They include primitive horses and camels, the oreodonts (an extinct even-toed ungulate group), and what appears to have been a particularly nasty representative of archaic, piglike beasts known as entelodonts. *Dinohyus,* the "terrible pig," was as big as a bison and had a huge head with a large mouth full of fearsome, bone-crunching teeth. It would undoubtedly have been a very effective scavenger. A site about a mile from the two hills is called Stenomylus Quarry for the unusual but abundant camel fossils there.

Near the two hills, in slightly younger sediments, a concentration of fossil doglike carnivores was excavated by the Carnegie crews in the early twentieth century and then reinvestigated by Bob Hunt from the University of Nebraska State Museum beginning in the 1980s. The disposition of the fossils indicates they were buried in a den. The carnivores dug dens and raised their young in the valley walls overlooking the ancient river, undoubtedly an opportune location for hunting prey drawn to the water. Most of the denning carnivores were bear-dogs, amphicyonids. Although doglike and about the size and proportion of wolves, they had diverged from the lineage of dogs long ago in their evolutionary history.

About a mile north of the entrance to the Agate Fossil Beds National Monument, excellent exposures of the Harrison formation close to Nebraska 29 provide a good look at one of the massive ancient soils that are characteristic of these fine-grained dune deposits. Look closely to see the snarl of roots and burrows that are preserved as small tubes and irregular shapes of calcium carbonate.

A couple miles farther north, Nebraska 29 climbs out of the Niobrara River valley and onto a low-relief surface developed on the Harrison formation. The road crosses this rolling grassland all the way to the town of Harrison. About 10 miles south of Harrison, a historic marker shows where the road crosses a nineteenth-century trade route from Fort Robinson to Fort Laramie, two frontier outposts in a wild and dangerous country.

The road that continues north from Harrison passes some interesting geology but becomes less developed as it goes. The area around Harrison is open with little relief and few exposures. About 1.5 miles north of town, a low outcrop along the road exposes two thin layers of volcanic ash in the Harrison formation. About 2 miles farther north, a small outcrop on the east side of the road reveals a cross section of a rooted and burrowed soil and, in the midst of it is a devil's corkscrew, one of the helical burrows of the rodent *Paleocastor*. A short distance farther north, the road follows the valley of Monroe Creek down the Pine Ridge escarpment, dropping through the Arikaree group to the White River group. This is the area from which the Monroe Creek formation of the Arikaree group takes its name. As throughout its entire extent, the Pine Ridge escarpment is created by erosion of the relatively resistant ledges of the Arikaree group. As the road continues out onto the flats north of Pine Ridge, small hills and buttes provide excellent exposures of the upper part of the White River group. Dirt roads continue north across the Oglala National Grassland to South Dakota.

Nebraska 71
Kimball—Scottsbluff
45 miles

Many of the geographical features in the Nebraska Panhandle run east-west, so north-south roads cross them and Nebraska 71 is no exception. From I-80, Nebraska 71 heads north across Lodgepole Creek and onto the Gangplank surface of Miocene age. It then crosses Quaternary-age channels, Pumpkin Creek, and Wildcat Ridge before reaching Scottsbluff on the North Platte River. It also takes you to an interesting geolocality—Scotts Bluff National Monument.

Nebraska 71 crosses Interstate 80 at exit 20 and heads north through Kimball. The town sits on a terrace of Lodgepole Creek, the largest stream that cuts into the Gangplank surface. This once-widespread, late Miocene surface is the top of the Ash Hollow formation of the Ogallala group. Where Lodgepole Creek and its tributaries have cut into the surface, the Ash Hollow formation is exposed with its characteristic siliceous mortar beds. There are also resistant calcareous ledges produced by soils that developed in the sands during periods of stability when the ground surface was neither eroding nor accumulating much additional sediment. The small drainages on the surface have not cut very deeply into the Ogallala sediment, and low, rolling hills characterize the landscape.

About 16 miles north of Kimball, you can see oil pumps to the west of the road. The southern part of the Nebraska Panhandle is on the edge of a deep structural basin, the Denver-Julesburg Basin, most of which is within Colorado. Cretaceous sandstones deeply buried beneath marine shales serve as a petroleum reservoir in this basin, which is the source for most of the meager petroleum production in Nebraska.

Another 6 miles north of the oil pumps (about 22 miles from Kimball), look for a gravel pit on the east side of Nebraska 71 where it begins to descend into the valley of Pumpkin Creek. The coarse conglomerate excavated from the quarry is poorly consolidated. Some of the last, and thus youngest, Ogallala sediment deposited, it fills a channel that cut down into older Ash Hollow sediments. During middle to late Miocene time when the Ogallala sediments were deposited, episodes of erosion scoured out valleys and then periods of deposition filled them with sand and gravel—a cycle that repeated itself many times.

Pumpkin Creek has cut its valley in older Tertiary sediments. Along the south valley wall, Ogallala sediments rest directly on the Oligocene Brule formation of the White River group, which shows up here as a brown siltstone. It appears that the Arikaree group sediments were completely stripped away on this side of the valley before deposition of the Ogallala group, but Arikaree sediments occur in a few places, sandwiched between the Brule and the Ogallala.

Geology along Nebraska 71 between Kimball and Scottsbluff. —*Modified from Swinehart and Diffendal, 1997; Burchett and Pabian, 1991*

Across the broad valley, you can see the north valley wall, a line of steep exposures known as Wildcat Ridge. Unlike the southern margin of the valley, which verges on a tableland, Wildcat Ridge really is a long, narrow ridge—a sharp divide between the Pumpkin Creek drainage and the valley of the North Platte River just to the north. Also unlike the southern valley wall is the sequence of sediments exposed along Wildcat Ridge. You can see siltstones of the Brule formation all along the lower slopes, and they undoubtedly underlie the Quaternary alluvium of the valley floor, but above the Brule formation is a thick sequence of Arikaree sediments. Only a few scattered remnants of Ogallala group sediments are preserved on the very top of the ridge.

The modern valley of Pumpkin Creek is just the last in a series of incisions that began in early Pleistocene time. Sands and gravels that were deposited in former channels are preserved at various heights along the south side of the valley with older channels higher than more recent deposits. The gravel in these Pleistocene channels includes abundant pebbles of anorthosite, an uncommon igneous rock that is made almost entirely of the mineral plagioclase, which is more susceptible to weathering than other common minerals such as potassium feldspar or quartz. The abundance of anorthosite means that these gravels have not been here long and did not travel far. They weathered out of a body of anorthosite, were transported some distance, and were deposited. They probably did not undergo many cycles of erosion and redeposition. The only likely source for anorthosite is in the Laramie Range in Wyoming, not very far west of here. This gives us an idea of the path they must have followed to get here—a path not very different from the modern North Platte drainage.

Just south of the intersection with Nebraska 88, a low outcrop by Nebraska 71 exposes some of the anorthosite-rich gravel deposited in Quaternary channels that were oriented almost directly east. The anorthosite pebbles are dark gray and coarse grained.

Nebraska 88 travels east, down Pumpkin Creek valley and beyond the end of Wildcat Ridge. Then the road and creek turn north, passing Courthouse Rock and Jail Rock and heading toward the North Platte River. One of the things you will notice about Pumpkin Creek, whether you follow it along Nebraska 88 or cross it on Nebraska 71, is how drastically underfit this stream is to the wide valley it flows down. Pumpkin Creek is just a trickle of water, a mere drainage ditch. The valley was probably carved out when much more water flowed in the creek than has in modern times, but the creek is even smaller now than it was only a few decades ago. Ranchers who divert surface water from the creek for agriculture have suggested that the increased withdrawals of groundwater from wells in the alluvium of the valley have decreased the creek flow. In an unconfined surface aquifer, such as the alluvium here is, surface waters are directly connected to the water in the ground. But a ruling by the Nebraska Supreme Court in 1966, based on an old state law that has since been overturned, made Nebraska the only western state that denies a connection between surface water and groundwater subflow when resolving legal disputes over water rights. The low water flow in this area, a serious situation for local agriculture, has developed into a court case that tests this ruling.

About 10 miles north of the junction with Nebraska 88, Nebraska 71 passes an old missile silo. These relics of the Cold War are

Geologists examine Quaternary gravels in a gravel pit in the Pumpkin Creek valley. Note the scour and fill that imposes one channel on another.

scattered throughout northern and western states within lobbing distance of the former Soviet Union. The United States military is decommissioning and dismantling many of these artifacts of technological destruction. Without the continued maintenance they received when operational, they become ever more vulnerable to the elements and more likely to interact with the environment. An empty silo is a deep hole in the ground that may contain hazardous materials since environmental concerns were not a priority when silos were designed and built.

As you approach Wildcat Ridge, buttes and monuments carved from the White River siltstones stand out from the ridge. As Nebraska 71 climbs up and over the ridge, it passes through a thick section of Arikaree sediments. Roadcuts on both sides of the ridge expose the Monroe Creek formation with its characteristic pipelike concretions. On the north side, the thin, horizontally bedded sands of the Gering formation are exposed below the Monroe Creek formation. As on Pine Ridge farther north, the Arikaree sediments support pine trees. At the top of the ridge is Wildcat Hills State Recreation Area, which boasts scenic views of the valleys to the north and south and provides campsites in the pines.

North of Wildcat Ridge, the road crosses a level area floored by surficial deposits that lie on top of the White River group. To the west the buttes of White River sediments capped by Arikaree sediments form a broken wall in the distance with Scotts Bluff National Monument at its north end. Nebraska 71 passes through the town of Gering, namesake of the Gering formation, crosses the North Platte, and enters the town of Scottsbluff.

Scotts Bluff National Monument

Scotts Bluff National Monument, just south of the town of Scottsbluff and west of Gering, preserves an important historical and geological site. As people migrated westward across America in the middle to late nineteenth century, they crossed what they called the Great American Desert, the area we now refer to as the Great Plains. The Oregon and Mormon Trails followed the broad Platte River floodplain across western Nebraska. The skyline of eroded monuments along the north branch of the Platte River provided welcome landmarks along the relatively featureless trails. But, even as they gave reassurance to the travelers, the hills and badlands of Scotts Bluff also proved an obstacle. The Mormon Trail kept to the north bank of the Platte, where the town now resides, but the Oregon Trail remained south of the massive cliff, following the same path as present-day Nebraska 92 and cutting through Mitchell Pass. You can still see the wagon ruts, deepened by erosion, at the national monument.

The entrance to the national monument is on the north side of Nebraska 92, just east of Mitchell Pass. Exhibits at the visitor center interpret the history of the area and provide some information about the geology. A winding road ascends Scotts Bluff to a small parking area at the top. The road climbs up through the upper part of the Brule formation of the White River group, with its pink and tan volcanic-rich siltstones. The White River group is the same sequence of late Eocene and Oligocene sediments that are exposed in badlands farther north at Toadstool Geologic Park and at Big Badlands National Park in South Dakota.

From the road, you can see two distinctive white beds of volcanic ash in the White River group. The upper ash is up to 10 feet thick in places. Although such local thicknesses could represent areas where the ash filled topographic depressions, it still represents a tremendous quantity of ash falling on this area. Some 30 million years ago, the nearest volcanoes were hundreds of miles to the west, so the eruptions that produced this ash must have been truly monstrous. Not far above this ash, the road passes upward through the evenly bedded sandstone of the Gering formation and the Monroe Creek formation that make up the Arikaree group.

At the end of the drive is a parking area, and a short walk takes you to an overlook with scenic views unmatched in the area. On a clear day, you can see many miles to the north and west. From the summit, the trail descends along the cliff face through the rock units and ends at the visitor center. The trail permits a close look at the sediments and their features.

View from atop Scotts Bluff looking to the southwest. The road from the visitor center winds up the hill to the top of the butte. The ledgy sandstone of the Arikaree group in the immediate foreground rests atop the massive White River siltstone. The pine-covered escarpment in the distance is Wildcat Ridge.

Pipelike concretions in the Monroe Creek formation are abundant on the trail from the parking lot to the overlook. In rocks at Pine Ridge farther north, these large elongate concretions were once used to distinguish the older Monroe Creek formation from the younger, concretionless Harrison formation. However, this difference is not consistent from place to place. One reason for this inconsistency is that the concretions did not form when the sediments were deposited. Instead, they grew within the sediment some time after deposition, by precipitation of calcium carbonate from the groundwater. The surface topography or the level of the water table may have controlled when and where the concretions formed. One peculiarity of these pipelike concretions is that their long axes are consistently aligned in a northeast-southwest orientation. They may be oriented in the direction of regional groundwater flow, but relatively little research has been done on this.

Also exposed along this upper part of the trail are pink or white lenses or layers of volcanic ash within dramatically crossbedded sandstones of the Monroe Creek sediments. Volcanic ash is also mixed in with the sand that encloses the ash beds. Crossbedded sandstones such as those exposed along the trail's concrete steps were once windblown sand dunes. Several characteristics distinguish these crossbeds from those that form in waterlain deposits. Dune sands tend to have large, extensive crossbeds, and fine layers are inversely graded—sand grains become progressively coarser from the bottom to the top of each fine layer. In normally graded sediment deposited in water, the coarser, heavier sand settles out first so it is at the base of the layer. In eolian sediments, the small layers near the base of the dune accumulate by a flow of loose grains that carries the coarser grains on top of turbulently moving, finer grains. You can see the inverse grading best in places where the eroded surface of the sandstone cuts across the fine layers at a low angle, making them appear very wide. The gradation of colors in each layer reflects the gradation of particle size in the sand grains. Only the lower part of the dunes is preserved.

The trail crosses over to the northeast side of the bluff and descends by switchbacks along the cliff. The finely laminated sandstones that characterize the Gering formation are well exposed along the north side of the ridge. Look for broad U-shaped downward depressions of the layers a few inches across. These may be cross-sectional views of footprints of ancient, hoofed mammals. You can also find crystals, seemingly composed of sand, in the top of the Gering formation. These crystals grew in and incorporated the sand grains. The cementing mineral matter is now calcite, but the form of the crystal indicates it was originally gypsum. Evaporation near the surface of the sands likely triggered the growth of gypsum. These conditions suggest a dry environment in keeping with the dune deposits of the Monroe Creek formation above. The Gering strata are thought to have been deposited by ephemeral streams.

Outcrop of Monroe Creek sandstones at the nose of Scotts Bluff where the trail first switches from the north side to the south side. At the base are poorly bedded sands with knobby potato-like concretions. Abundant burrowing may have destroyed original laminations. Note the white ash layer above the contact between the poorly bedded and well-bedded sands. In 2002 a good size rockfall closed the trail here for a while.

The trail passes through a human-made tunnel. A major ash layer forms the roof of the tunnel. The bottom surface of the ash has a lot of microrelief on it, and some of the cylindrical forms may be ash-filled burrows.

On the south side of the tunnel, you can see why the face of this cliff is relatively flat. A suite of joints runs parallel to the face. Note the slabs of the sedimentary rock tilting away from and sliding down along the cliff face. In 1996 a slab of rock on this side of the cliff gave way, producing a good-sized rock fall. Fresh rock debris mantles the talus slope above the trail. The joint direction, which is parallel to the North Platte River valley and the long axis of Wildcat Ridge, likely controlled drainage development.

Farther down the trail, on the grassy flats, if you look back to the north at the cliff you can see the strange, undulating surface that separates the overlying Gering formation of the Arikaree group from the Brule formation of the White River group. Geologists don't know if erosion or other processes formed this peculiar surface before the Gering formation was deposited, or if the still-unconsolidated sediments flowed after burial in response to loading or seismic vibrations. In other parts of the national monument, the contact is planar and flat.

Along the lower part of the trail, grass covers the Brule formation. Water following fractures in the relatively impermeable siltstones of the formation emerges as a spring, a common occurrence throughout the formation.

An erosional fin of Scotts Bluff viewed from the trail. The contorted surface near the base of the cliff *(inset)* is the contact between the Brule formation of the White River group below and the Gering formation of the Arikaree group above. The thin white layer in the Gering sediment midway up the bluff is a volcanic ash bed.

GLOSSARY

anorthosite. An igneous rock composed mostly of the mineral plagioclase. Uncommon at the earth's surface.

aquiclude. A body of rock that does not allow the passage of groundwater.

aquifer. A body of rock with enough open spaces (pores) to hold substantial amounts of groundwater. The pores are connected, allowing the transmission of the water, for example, into a well.

arch. A broad upwarping of the crust that occurs over a wide area (hundreds of square miles).

badlands. An area of relatively soft sediments that have been extensively dissected into narrow canyons with steep walls. They usually develop in dry climates.

basalt. A dark volcanic rock rich in iron and magnesium. It is usually associated with non-explosive eruptions of lava.

bioturbation. The disturbance of sedimentary layering by extensive burrowing.

calcareous. Containing calcium carbonate as in sediment or sedimentary rock.

carbonate. A general name given to sedimentary rocks that are composed of carbonate minerals; either limestone or dolostone.

Cenozoic. Era of geologic time, sometimes referred to as the "Age of Mammals," that extended from about 65 million years ago to the present.

chert. A fine-grained sedimentary rock composed of microscopic quartz.

clast. A particle in a sedimentary deposit.

coal. A sedimentary rock made by the accumulation and anaerobic decomposition of plant matter in a quiet environment. It can be mined and burned as an energy resource.

concretion. Sedimentary structures formed after deposition by greater cementation of small areas within the sediment and later exposed by the weathering and erosion of the softer surrounding sediment. Sometimes organic material creates the local chemical environment that causes this selective cementation.

conglomerate. A coarse-grained sedimentary rock that is composed of a wide variety of grain sizes. It forms as a result of cementation of rounded gravel.

cyclothem. A sequence of sedimentary rocks deposited during cyclical rises and falls of sea level. Cyclothems are common in late Paleozoic rocks in the midcontinent area of North America.

debris flow. A mass wasting event, in which a chaotic mixture of weathered material moves and is deposited as a slurry.

247

disconformity. An unconformity in which the strata above and below the depositional contact are parallel to each other.

dolostone. A sedimentary rock composed mostly of the mineral dolomite, a calcium-magnesium carbonate mineral. It is very similar to limestone.

entrenched stream. A stream that is actively downcutting through its recent floodplain. It has not had time to establish a new equilibrium and form a new floodplain at the lower level.

eolian deposits. Sand and silt deposits that were transported by wind.

erratic. A large rock transported and deposited by a glacier. It is commonly of a different composition than bedrock in the local area.

evaporite. Sedimentary deposit of minerals formed by precipitation from water solution that has been concentrated by evaporation in a dry climate. Gypsum is a commn evaporite formed from the evaporation of seawater.

fault. A break in the earth's crust along which movement has taken place.

floodplain. The level area alongside a river that is prone to periodic flooding. It forms by both erosional and depositional processes.

gneiss. A banded metamorphic rock formed under high pressures and temperatures. Segregation of different minerals in different layers forms the bands.

granite. An igneous rock that forms from the slow cooling of molten magma within the earth's crust. It is composed primarily of the minerals quartz, feldspar, and mica.

gravel. Sedimentary particles, larger than 2 millimeters in diameter.

greenstone. A green metamorphic rock, formed by metamorphism of the volcanic rock basalt at relatively low temperatures and pressures. The characteristic green metamorphic minerals in the rock form at these low temperatures and pressures.

lens. A thin discontinuous sedimentary deposit that pinches out laterally.

limestone. A sedimentary rock composed of the mineral calcite (calcium carbonate). It forms in warm marine environments and commonly contains fossils of shells.

loess. A deposit of weakly cemented silt that was deposited by the wind. Loess can form steep cliff faces, although the material has low strength.

marl. A name given to impure limestones and dolostones. The contaminants are usually sand or clay.

mass wasting. The downslope movement of soil, rocks, or rock debris under the influence of gravity and without the aid of a flowing medium such as water.

Mesozoic. Era of geologic time, popularly called the "Age of the Dinosaurs," that lasted from approximately 250 to 65 million years ago.

moraine. A ridge of sediment deposited where a glacier is actively melting. Composed of till, a sediment that contains particles of a wide variety of sizes.

orogeny. The various tectonic processes that collectively lead to mountain building.

paleosoil (paleosol). An ancient soil layer preserved in a sequence of sediments or sedimentary rocks. Its presence indicates that the sediments were exposed at the surface long enough for soil to form. It gives important clues about the climate it formed in.

Paleozoic. Era of geologic time that lasted from approximately 545 to 250 million years ago.

Pangaea. The name given to an ancient supercontinent that consisted of all of the major continental regions. It existed at the end of the Paleozoic Era.

point bar. A sandbar that forms on the inside of the bend of a meandering river. As the point bar is added to and the bank on the outside of the bend is eroded, the channel shifts laterally.

Precambrian. That part of the geologic record that is older than 545 million years. This era of time constitutes moe than seven-eighths of the earth's history.

quartzite. A metamorphic rock consisting of the mineral quartz. Before metamorphism, it was probably a quartz sandstone.

Quaternary. Period of geologic time extending from approximately 1.8 million years ago to today.

redbeds. Sedimentary rocks that are red, usually because of the presence of iron-oxide cement holding the sediments together. They form in a terrestrial environment.

rhyolite. A volcanic rock rich in silicon and aluminum. It commonly is expelled from a volcano during extremely violent eruptions.

rift. A linear geologic feature that forms when the crust is pulled apart. Long faults form, and a block of the crust drops down between them under the influence of gravity, forming a long linear valley. Eruptions of basalt and rhyolite often accompany rift formation.

riparian. Along a river; the wooded environments close to a river in an otherwise dry climate.

sandstone. A sedimentary rock composed of cemented sand grains, usually composed of the mineral quartz.

shale. A sedimentary rock composed of mud deposited in a quiet, low-energy environment.

siliceous tests. Microscopic shells, mainly of diatoms, that are composed of silica-rich material. These accumulate when the animals die and the shells sink to the basin floor.

stromatolites. Sedimentary structures formed by calcareous sediments trapped by algal filaments. They often form columnar structures with concave upwards laminations.

terrace. A flat area, above the active floodplain of a river, that represents the remains of an older, higher floodplain. Terraces form when streams undergo accelerated erosion to a new, lower level.

Tertiary. Period of geologic time that extends from approximately 65 to 1.8 million years ago.

till. An accumulation of sediment deposited by a melting glacier. It is usually composed of a wide range of sediment sizes.

type section. A formally defined, local stratigraphic section that serves as a reference point for a particular formation or unit usually where it was originally described.

unconformity. A break in a sequence of sedimentary rocks representing a significant interval of geologic time, usually caused by a period of erosion.

ungulate. Any hoofed mammal.

volcanic ash. Very fine-grained particles produced during violent volcanic eruptions. Wind can transport ash great distances before deposition.

water table. The top of the zone of groundwater saturation.

Suggested Reading and Further References

Asterisks mark references that we recommend for the layperson

Agenbroad, L. D. 1978. *The Hudson-Meng Site: An Alberta Bison Kill in the Nebraska High Plains*. Caldwell, Idaho: The Caxton Printers, Ltd.

Anonymous. 1993. Ancient Lake Diffendal created by dune dam on North Platte. *Resource News* II(1). Lincoln: University of Nebraska Conservation and Survey Division.

*Bleed, A., and C. Flowerday. 1989. *An Atlas of the Sand Hills*. Lincoln: University of Nebraska Conservation and Survey Division, Institute of Agriculture and Natural Resources.

Burchett, R. R. 1987. Pennsylvanian and Permian rocks associated with the Humboldt Fault Zone in southeastern Nebraska. *Geological Society of America Centennial Field Guide—North Central Section*, 35–38. Boulder, Colorado: Geological Society of America.

Burchett, R. R., V. H. Dreeszen, E. C. Reed, and G. E. Prichard. 1972. *Bedrock Geologic Map Showing Thickness of Overlying Quaternary Deposits, Lincoln Quadrangle and Part of Nebraska City Quadrangle, Nebraska and Kansas*. U.S. Geological Survey Miscellaneous Investigations Series Map I-729.

Burchett, R. R., V. H. Dreeszen, V. L. Souders, and G. E. Prichard. 1988. *Bedrock Geologic Map Showing Configuration of the Bedrock Surface in the Nebraska Part of the Sioux City 1°x2° Quadrangle*. U.S. Geological Survey Miscellaneous Investigations Series Map I-1879.

Burchett, R. R., and R. K. Pabian. 1991. *Geologic Bedrock Map of Nebraska*. Lincoln: University of Nebraska Conservation and Survey Division, Institute of Agriculture and Natural Resources.

Burchett, R. R., E. C. Reed, and V. H. Dreeszen. 1975. *Bedrock Geologic Map Showing Thickness of Overlying Quaternary Deposits, Fremont Quadrangle and Part of Omaha Quadrangle, Nebraska*. U.S. Geological Survey Miscellaneous Investigations Series Map I-905.

Carlson, M. P. 2000. Island arcs, accretionary terranes and midcontinent structure: New understandings of the geologic architecture of the U.S. midcontinent. *Resource Notes* 15(1):15–18. Lincoln: University of Nebraska Conservation and Survey Division.

*Carlson, M. P. 1993. *Geology, Geologic Time and Nebraska*. Lincoln: University of Nebraska Conservation and Survey Division, Institute of Agriculture and Natural Resources.

*The Cellars of Time. 1994. *Nebraskaland Magazine.* Nebraska Game and Parks Commission.

Dethier, D. P. 2001. Pleistocene incision rates in the western United States calibrated using Lava Creek B tephra. *Geology* 29:783–86.

251

Diffendal, R. F., Jr. 1991. *Geologic Map Showing Configuration of the Bedrock Surface, North Platte 1°x2° Quadrangle, Nebraska.* U.S. Geological Survey Miscellaneous Investigations Series Map I-2277.

Diffendal, R. F., Jr. 1987. Ash Hollow State Historical Park: Type area for the Ash Hollow Formation (Miocene), western Nebraska. *Geological Society of America Centennial Field Guide—North Central Section,* 29–34. Boulder, Colorado: Geological Society of America.

Diffendal, R. F., Jr., and C. Flowerday. 1995. *Geologic Field Trips in Nebraska and Adjacent Parts of Kansas and South Dakota.* Lincoln: University of Nebraska Conservation and Survey Division, Institute of Agriculture and Natural Resources.

*Diffendal, R. F., Jr., R. K. Pabian, and J. R. Thomason. 1969. *Geologic History of Ash Hollow Park Nebraska.* Lincoln: University of Nebraska Conservation and Survey Division, Institute of Agriculture and Natural Resources.

Diffendal, R. F., Jr., and M. R. Voorhies. 1994. *Geologic Framework of the Niobrara River Drainage Basin and Adjacent Areas in South Dakota Generally East of the 100th Meridian West Longitude and West of the Missouri River.* Lincoln: University of Nebraska Conservation and Survey Division, Institute of Agriculture and Natural Resources.

Dreeszen, V. H., E. C. Reed, R. R. Burchett, and G. E. Prichard. 1973. *Bedrock Geologic Map Showing Thickness of Overlying Quaternary Deposits, Grand Island Quadrangle, Nebraska and Kansas.* U.S. Geological Survey Miscellaneous Investigations Series Map I-819.

Eversoll, D. A., V. H. Dreeszen, R. R. Burchett, and G. E. Prichard. 1988. *Bedrock Geologic Map Showing the Configuration of the Bedrock Surface, McCook 1°x2° Quadrangle, Nebraska and Kansas, and Part of the Sterling 1°x2° Quardrangle, Nebraska and Colorado.* U.S. Geological Survey Miscellaneous Investigations Series Map I-1878.

Flowerday, C. A., and R. F. Diffendal, Jr. 1997. *Geology of Niobrara State Park, Knox County, Nebraska, and Adjacent Areas—with a Brief History of the Park, Gavins Point Dam, and Lewis and Clark Lake.* Lincoln: University of Nebraska Conservation and Survey Division, Institute of Agriculture and Natural Resources.

Galusha, T. 1975. Stratigraphy of the Box Butte formation, Nebraska. *Bulletin of the American Museum of Natural History* 156:5–67.

*Loope, D. B., and J. B. Swinehart. 2000. Thinking like a dune field: geologic history in the Nebraska Sand Hills. *Great Plains Research* 10(1):5–35

Mandel, R. D., and E. A. Bettis. 2000. *Late Quaternary Landscape Evolution in the South Fork of the Big Nemaha River Valley, Southeastern Nebraska and Northeastern Kansas.* Lincoln: University of Nebraska Conservation and Survey Division, Institute of Agriculture and Natural Resources.

Martin, J. E. 1985. Fossiliferous Cenozoic deposits of western South Dakota and northwestern Nebraska. *Dakoterra* 2(2):7. Rapid City: South Dakota School of Mines and Technology, Museum of Geology.

McKee, E. D. 1979. *A Study of Global Sand Seas.* U.S. Geological Survey Professional Paper 1052.

*Pabian, R. K. 1987. *Geology of Selected Sites near the Republican River in Franklin County, Nebraska.* Lincoln: University of Nebraska Conservation and Survey Division, Institute of Agriculture and Natural Resources.

Pabian, R. K. 1987. The Late Cretaceous Niobrara formation in south central Nebraska. *Geological Society of America Centennial Field Guide—North Central Section,* 39–42. Boulder, Colorado: Geological Society of America.

Pabian, R. K., and R. F. Diffendal, Jr. 1991. *Late Paleozoic Cyclic Sedimentation in Southeastern Nebraska: A Field Guide.* Lincoln: University of Nebraska Conservation and Survey Division, Institute of Agriculture and Natural Resources.

*Pabian, R. K., and R. F. Diffendal, Jr. 1969. *Geology of Lake McConaughy Area, Keith County, Nebraska.* Lincoln: University of Nebraska Conservation and Survey Division, Institute of Agriculture and Natural Resources.

Pabian, R. K., and D. R. Lawton. 1987. Late Cretaceous strata exposed in Ponca State Park. *Geological Society of America Centennial Field Guide—North Central Section,* 7–13. Boulder, Colorado: Geological Society of America.

*Pabian, R. K., and J. B. Swinehart. 1979. *Geologic History of Scotts Bluff National Monument.* Lincoln: University of Nebraska Conservation and Survey Division, Institute of Agriculture and Natural Resources.

Skinner, M. F. 1972. Early Pleistocene glacial and pre-glacial rocks and faunas of north-central Nebraska. *Bulletin of the American Museum of Natural History* 148:1–148.

Skinner, M. F., S. M. Skinner, and R. J. Gooris. 1977. Statigraphy and biostratigraphy of late Cenozoic deposits in central Sioux County, western Nebraska. *Bulletin of the American Museum of Natural History* 158:263–370.

Souders, V. L., J. B. Swinehart, and V. H. Dreeszen. 1990. *Postulated Evolution of the Platte River and Related Drainages, 1.* Lincoln: University of Nebraska Conservation and Survey Division, Institute of Agriculture and Natural Resources.

Swinehart, J. B. 1989. Wind-blown Deposits. In *An Atlas of the Sand Hills.* Ed. A. Bleed and C. Flowerday, 43–56. Lincoln: University of Nebraska Conservation and Survey Division.

Swinehart, J. B., and R. F. Diffendal, Jr. 1997. *Geologic Map of the Scottsbluff 1°x2° Quadrangle, Nebraska and Colorado.* U.S. Geological Survey Miscellaneous Investigations Series Map I-2545.

Swinehart J. B., and R. F. Diffendal, Jr. 1995. *Geologic Map of Morrill County, Nebraska.* U.S. Geological Survey Miscellaneous Investigations Series Map I-2496.

Swinehart, J. B., and R. F. Diffendal, Jr. 1987. Duer Ranch, Morrill County, Nebraska: Contrast between Cenozoic fluvial and eolian deposition. *Geological Society of America Centennial Field Guide—North Central Section,* 23–28. Boulder Colorado: Geological Society of America.

Swinehart, J. B., V. H. Dreeszen, G. M. Richmond, M. J. Bretz, F. V. Steece, G. R. Hallberg, and J. E. Goebel. 1994. *Quaternary Geologic Map of the Platte River 4°x6° Quadrangle, United States.* U.S. Geological Survey Miscellaneous Geologic Investigations Map I-1420.

Swinehart, J. B., and D. Loope. 1987. Late Cenozoic geology along the summit to museum hiking trail, Scotts Bluff National Monument, western Nebraska. *Geological Society of America Centennial Field Guide—North Central Section,* 13–18. Boulder, Colorado: Geological Society of America.

Swinehart, J. B., V. L. Souders, H. M. DeGraw, and R. F. Diffendal, Jr. 1985. Cenozoic Paleogeography of Western Nebraska. In *Cenozoic Paleogeography of West-Central United States, Rocky Mountain Section.* Ed. R. M. Flores and S. S. Kaplan, 209–29. Denver: Society of Economic Paleontologists and Mineralogists.

Van Schmus, R., and others. 1993. Transcontinental Proterozoic Provinces. In *Precambrian Conterminous U.S.: The Geology of North America, C-2.* Ed. J. C. Reed and others, 171–334. Boulder, Colorado: Geological Society of America.

Voorhies, M. R. 1987. Late Cenozoic stratigraphy and geomorphology, Fort Niobrara, Nebraska. *Geological Society of America Centennial Field Guide—North Central Section,* 1–6. Boulder, Colorado: Geological Society of America.

Wayne, W. J. 1987. The Platte River and Todd Valley, near Fremont, Nebraska. *Geological Society of America Centennial Field Guide—North Central Section,* 19–22. Boulder, Colorado: Geological Society of America.

Websites with Information on Nebraska's Geology

The Web is so dynamic, with servers, addresses, and content changing on a regular basis, that a list of specific URLs for specific information will likely not be very useful several years after this is written. For this reason, we have compiled a partial list of organizations and their homepages that support websites with information on Nebraska's geology.

Center for Advanced Land Management Information Technologies at University of Nebraska at Lincoln
http://www.calmit.unl.edu/

A variety of more technical information and data is available here, including Geographic Information System files in ArcInfo and ArcView format.

Geological Society of America
http://www.geosociety.org/index.htm

This is an extensive site dedicated to one of the larger geoscience professional organizations, with some information on Nebraska's geology buried here and there within it. Some of the information is fairly technical. It does provide introductory material and information on geology in general, and has an internal search engine.

NASA
http://www.nasa.gov/

Buried in this site at a number of places are a variety of types of imagery of Nebraska that you can download, including satellite images and images taken by astronauts.

National Park Service
Agate Fossil Beds National Monument
http://www.aqd.nps.gov/grd/parks/agfo/

Scotts Bluff National Monument
http://www.nps.gov/scbl/

The National Park Service generally provides information for the layperson about the geology of their various parks.

Nebraska Conservation and Survey Division
http://csd.unl.edu/csd.html

This is a good place to start when searching for information on Nebraska's geology. It provides a list of publications, links to other state sites that have geologic information, educational material, and much more.

Nebraska Department of Natural Resources
http://www.nrc.state.ne.us/

This site has a lot of information on surface and groundwater, and it also contains a digital elevation model (DEM) databank.

Nebraska Game and Parks Commission
http://www.ngpc.state.ne.us/homepage.html

We included this because many of the geolocalities and best places for public access to the land and geology are state parks, and this site provides general information on these.

TerraServer Repository
http://terraserver.homeadvisor.msn.com/default.asp

This is good site for acquiring U.S. Geological Survey air photo imagery of almost any part of Nebraska.

United States Geological Survey
http://www.usgs.gov/

Information about Nebraska is scattered in a variety of places in this very large site. There is also extensive educational material for the layperson on geology in general.

University of Nebraska State Museum
http://www.museum.unl.edu/index.html

This is the website for Morrill Hall, the state museum with extensive displays of Nebraska fossils and one of the geolocalities in this book.

INDEX

George F. Engelmann, Harmon D. Maher, Jr., and Robert D. Shuster

About the Authors

For more than fifteen years, **Harmon D. Maher, Jr., George F. Engelmann,** and **Robert D. Shuster** have taught geology at the University of Nebraska at Omaha. While taking students on field trips across the state, they wished they had a book describing Nebraska's geologic features, many of which are world-class.

When he is not teaching, kayaking, or spending time with his family, Harmon Maher studies the tectonic history of Svalbard, a group of islands in the Arctic Ocean. George Engelmann specializes in vertebrate paleontology—specifically mammal fossils and their distribution through time and place—something of particular interest in Nebraska. Robert Shuster, the geochemist in the trio, researches such diverse topics as environmental geology, geoarchaeology, and the origin of granites.

We encourage you to patronize your local bookstore. Most stores will order any title they do not stock. You may also order directly from Mountain Press, using the order form provided below or by calling our toll-free, 24-hour number and using your VISA, MasterCard, Discover or American Express.

Some geology titles of interest:

____ROADSIDE GEOLOGY OF ALASKA	18.00
____ROADSIDE GEOLOGY OF ARIZONA	18.00
____ROADSIDE GEOLOGY OF COLORADO, 2nd Edition	20.00
____ROADSIDE GEOLOGY OF HAWAII	20.00
____ROADSIDE GEOLOGY OF IDAHO	20.00
____ROADSIDE GEOLOGY OF INDIANA	18.00
____ROADSIDE GEOLOGY OF LOUISIANA	15.00
____ROADSIDE GEOLOGY OF MAINE	18.00
____ROADSIDE GEOLOGY OF MASSACHUSETTS	20.00
____ROADSIDE GEOLOGY OF MONTANA	20.00
____ROADSIDE GEOLOGY OF NEBRASKA	18.00
____ROADSIDE GEOLOGY OF NEW MEXICO	16.00
____ROADSIDE GEOLOGY OF NEW YORK	20.00
____ROADSIDE GEOLOGY OF NORTHERN and CENTRAL CALIFORNIA	20.00
____ROADSIDE GEOLOGY OF OREGON	16.00
____ROADSIDE GEOLOGY OF PENNSYLVANIA	20.00
____ROADSIDE GEOLOGY OF SOUTH DAKOTA	20.00
____ROADSIDE GEOLOGY OF TEXAS	20.00
____ROADSIDE GEOLOGY OF UTAH	18.00
____ROADSIDE GEOLOGY OF VERMONT & NEW HAMPSHIRE	14.00
____ROADSIDE GEOLOGY OF VIRGINIA	16.00
____ROADSIDE GEOLOGY OF WASHINGTON	18.00
____ROADSIDE GEOLOGY OF WYOMING	18.00
____ROADSIDE GEOLOGY OF THE YELLOWSTONE COUNTRY	12.00
____GLACIAL LAKE MISSOULA AND ITS HUMONGOUS FLOODS	15.00
____AGENTS OF CHAOS	14.00
____COLORADO ROCKHOUNDING	20.00
____NEW MEXICO ROCKHOUNDING	20.00
____FIRE MOUNTAINS OF THE WEST	18.00
____GEOLOGY UNDERFOOT IN CENTRAL NEVADA	16.00
____GEOLOGY UNDERFOOT IN DEATH VALLEY AND OWENS VALLEY	16.00
____GEOLOGY UNDERFOOT IN ILLINOIS	15.00
____GEOLOGY UNDERFOOT IN SOUTHERN CALIFORNIA	14.00
____NORTHWEST EXPOSURES	24.00

Please include $3.00 per order to cover postage and handling.

Send the books marked above. I enclose $_____

Name _____

Address _____

City/State/Zip _____

☐ Payment enclosed (check or money order in U.S. funds)

Bill my: ☐ VISA   ☐ MasterCard ☐ Discover ☐ American Express

Card No. _____ Expiration Date:_____

Signature _____

MOUNTAIN PRESS PUBLISHING COMPANY
P.O. Box 2399 • Missoula, MT 59806 • Order Toll-Free 1-800-234-5308
E-mail: info@mtnpress.com • Web: www.mountain-press.com